AS/A-LEVEL

STUDENT GUIDE

OCR

Geography

Investigative geography
Geographical and fieldwork skills

Peter Stiff and David Barker

HODDER EDUCATION
AN HACHETTE UK COMPANY

Hodder Education, an Hachette UK company, Blenheim Court, George Street, Banbury, Oxfordshire OX16 5BH

Orders

Bookpoint Ltd, 130 Milton Park, Abingdon, Oxfordshire OX14 4SB

tel: 01235 827720

fax: 01235 400401

e-mail: education@bookpoint.co.uk

Lines are open 9.00 a.m.–5.00 p.m., Monday to Saturday, with a 24-hour message answering service. You can also order through the Hodder Education website: www.hoddereducation.co.uk

ISBN 978-1-4718-6413-1

First printed 2017

Impression number 5 4 3 2 1

Year 2021 2020 2019 2018 2017

Cover photo: Richard Carey/Fotolia; photograph on page 71: David Bagnall/Alamy Stock Photo

Typeset by Integra Software Services Pvt Ltd, Pondicherry, India

Printed in Italy

Hachette UK's policy is to use papers that are natural, renewable and recyclable products and made from wood grown in sustainable forests. The logging and manufacturing processes are expected to conform to the environmental regulations of the country of origin.

Contents

Content Guidance

Questions & Answers

Getting the most from this book

Exam tips

Advice on key points in the text to help you learn and recall content, avoid pitfalls, and polish your exam technique in order to boost your grade.

Knowledge check

Rapid-fire questions throughout the Content Guidance section to check your understanding.

Knowledge check answers

1 Turn to the back of the book for the Knowledge check answers.

Summaries

- Each core topic is rounded off by a bullet-list summary for quick-check reference of what you need to know.

Exam-style questions

Commentary on the questions

Tips on what you need to do to gain full marks, indicated by the icon ⓔ

Sample student answers

Practise the questions, then look at the student answers that follow.

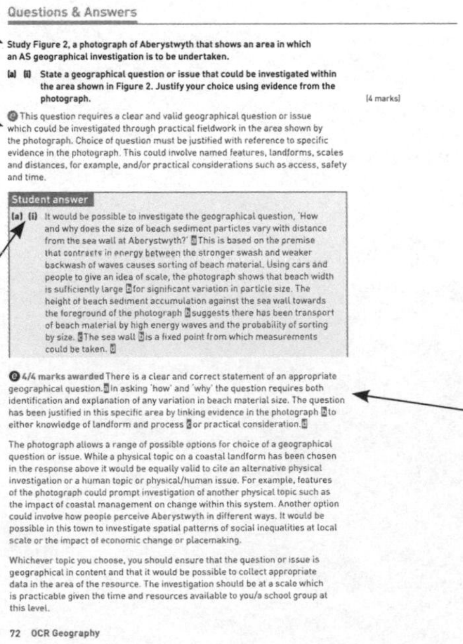

Questions & Answers

Study Figure 2, a photograph of Aberystwyth that shows an area in which an AS geographical investigation is to be undertaken.

(a) (i) State a geographical question or issue that could be investigated within the area shown in Figure 2. Justify your choice using evidence from the photograph. (4 marks)

ⓔ This question requires a clear and valid geographical question or issue which could be investigated through practical fieldwork in the area shown by the photograph. Choice of question must be justified with reference to specific evidence in the photograph. This could involve named features, landforms, scales and distances, for example, and/or practical considerations such as access, safety and time.

Student answer

(a) (i) It would be possible to investigate the geographical question, 'How and why does the size of beach sediment particles vary with distance from the sea wall at Aberystwyth?' [a] This is based on the premise that contrasts in energy between the stronger swash and weaker backwash of waves causes sorting of beach material. Using cars and people to give an idea of scale, the photograph shows that beach width is sufficiently large [b] for significant variation in particle size. The height of beach sediment accumulation against the sea wall towards the foreground of the photograph [b] suggests there has been transport of beach material by high energy waves and the probability of sorting by size. [c] The sea wall [b] is a fixed point from which measurements could be taken. [d]

ⓔ **4/4 marks awarded** There is a clear and correct statement of an appropriate geographical question. [a] In asking 'how' and 'why' the question requires both identification and explanation of any variation in beach material size. The question has been justified in this specific area by linking evidence in the photograph [b] to either knowledge of landform and process [c] or practical consideration. [d]

The photograph allows a range of possible options for choice of a geographical question or issue. While a physical topic on a coastal landform has been chosen in the response above it would be equally valid to cite an alternative physical investigation or a human topic or physical/human issue. For example, features of the photograph could prompt investigation of another physical topic such as the impact of coastal management on change within this system. Another option could involve how people perceive Aberystwyth in different ways. It would be possible in this town to investigate spatial patterns of social inequalities at local scale or the impact of economic change or placemaking.

Whichever topic you choose, you should ensure that the question or issue is geographical in content and that it would be possible to collect appropriate data in the area of the resource. The investigation should be at a scale which is practicable given the time and resources available to you/a school group at this level.

72 OCR Geography

Commentary on sample student answers

Read the comments (preceded by the icon ⓔ) showing how many marks each answer would be awarded in the exam and exactly where marks are gained or lost.

About this book

Success in A-level geography requires not only substantial knowledge and authoritative understanding of the specification content but also the development of geographical and study skills. Some of these skills, including fieldwork, will be acquired as you study the content. If you are studying for AS then two days of fieldwork are the minimum required. Fieldwork experience is extended to a minimum of four days at A-level. Fieldwork undertaken for the independent investigation may contribute to these four days.

Geographical skills should be developed throughout both AS and A-level courses, not as a separate theme or topic. These skills are fundamental to developing substantial knowledge and authoritative understanding of the geography studied. Your teachers will introduce the range of skills required as appropriate throughout the course. In this book, these skills can be found in the five chapters making up the Investigative geography section.

The *OCR Geography* textbook published by Hodder Education (978-1-4718-5870-3) contains helpful material on both geographical and fieldwork skills. In addition, *Essential Maths Skills for AS/A-level Geography* (978-1-4718-6355-4), also published by Hodder Education, will be useful.

At A-level, fieldwork skills are assessed within the Investigative geography component (04/05). Geographical skills are assessed within the three other components: Physical systems (01), Human interactions (02) and Geographical debates (03).

The **Investigative geography** component offers you the opportunity to undertake an independent investigation. There are 60 marks available and it contributes 20% of the total A-level qualification. This guide will help you develop a wide range of skills and will support you as you conduct your individual investigation.

The Investigative geography component has no prescribed geographical content but does have five key sections that your written report *must* contain:

- Planning, purpose and introduction
- Data, information collection methods and sampling framework
- Data presentation techniques
- Data analysis and explanation
- Conclusions and investigation evaluation
- Overall quality and communication of written work (this is not a separate section, but the whole of your report will be assessed under this heading)

This guide is structured using these sections as a framework. Each of these sections contain:

- the principal requirements
- how to achieve a high level of performance within them

The assessment of geographical skills at AS is within both Landscape and place (01) and Geographical debates (02). In Paper 01, Section C is headed 'Fieldwork' and is focused on assessing fieldwork skills.

This guide consists of:

Content Guidance: this section summarises some of the key information that you need to know to be able to conduct a successful investigation. It includes both geographical and fieldwork skills. In particular, the meaning of key terms is made clear and some attention is paid to providing details of where skills are appropriate when studying the content of the other components at both AS and A-level. Students will also benefit from noting the **Exam tips** that will provide further help in determining how to make the most of your investigation. **Knowledge check** questions are designed to help learners check their depth of knowledge — why not get someone else to ask you these!

Questions & Answers: this section includes sample questions similar in style to those you might expect in the AS exam papers. There are sample student responses to these questions as well as detailed analysis, which will give further guidance on what exam markers are looking for to award top marks.

Content Guidance

Investigative geography

Overview of the investigative geography component

Aims of the component

This component contributes to you becoming confident and competent in selecting, using and evaluating a range of quantitative and qualitative skills and in how to approach geographical investigation. It is also aimed at helping you understand the fundamental role of fieldwork as a tool to understand and generate new knowledge about the real world, in particular about the location(s) you personally investigate.

Additionally, tackling a self-directed investigation helps you develop skills in planning, undertaking and evaluating a more extended assignment. These skills are vital for success in higher education and are also directly transferrable to the world of work.

Your investigation must:

- be based on a question or issue defined and developed by you individually to address aims and questions relating to any of the specification content
- incorporate data and/or evidence from your own investigations, collected individually or in groups
- draw on your own research, including field data and secondary data sourced by you
- set your findings and data in their wider context and analyse and summarise them
- draw your own conclusions

Key features of your independent investigation

1. It must be a written report in continuous prose.
2. It must have a clear structure.
3. The recommended word length is between 3,000 and 4,000 words.
4. It must include the collection of **primary** and **secondary data** and/or evidence.
5. Digital material must be referenced in full and evidence such as web links and/or screenshots included in the report.

Primary data are unprocessed data such as those you collect in the field. Some published data are primary as they have not been worked on, such as being analysed and/or interpreted. Some census material, electoral rolls and remote-sensed data are examples.

Secondary data come from published documentary sources that have been analysed and/or interpreted. Research papers, published maps, textbooks are examples.

The level of independence required

Essentially, for the investigative geography component an individual student is expected to take the initiative and provide the energy and stamina to see the project through from start to finish. However, there are some aspects of the process where collaboration is allowed. The two levels of student independence are:

1 independent work: students must work alone
2 collaboration allowed: students may work as a class/group/pair

Working with others is not compulsory, so where collaboration is allowed you may choose to work alone. See Table 1 for a summary of which aspects allow collaboration and which require independent work.

Table 1 Levels of independence and investigation stages

Stage or aspect of investigation	Level of independence allowed	Example(s)
Exploring the focus	Collaboration	Students can be given a completely free choice of investigation or a Centre can suggest a theme or range of themes. Students may discuss ideas and research for appropriate questions
Choosing a title, the investigation's focus and purpose	Independent	Students must not be given a list of titles/investigations. Each student must finalise the title and justify how their enquiry will help them address their title and explore their theme in the context of the chosen geographical location
Devising methodology and sampling framework	Collaboration	Students may work together when planning and choosing methodologies and sampling strategies
Primary data collection	Collaboration	Students may collect primary data individually or in groups
Secondary data collection	Independent	Students must choose secondary data sources on their own and carry out data collection from these independently
Data/information presentation	Independent	Students must select and use data presentation methods on their own
Data analysis and explanation/ interpretation	Independent	Students must select and use analysis methods on their own and independently interpret and analyse the results
Conclusions and evaluation	Independent	Students must evaluate their investigation findings and reach a balanced and supported conclusion on their own

If an investigation involves data collection by a group, an individual student must describe their particular role in this process. When the report is submitted it must be accompanied by a Geography Independent Investigation Form. The student must sign a declaration to confirm that, apart from the allowed collaboration, the work is their own unaided work.

This form also has an 'Investigation proposal section' where a student enters their investigation title, hypotheses and/or questions and any possible subquestions, enquiry route and suggested data collection methods. A teacher is allowed to check this before the investigation proceeds. This is to make sure that the student's proposal is 'fit for purpose'. This review will ensure that the proposed investigation is suitable, accessible and achievable with the resources available. It also ensures that the assessment criteria by which the investigation will be marked are covered by the student's proposal.

Exam tip

If you are unsure about whether or not you have to carry out any aspect of your investigation individually, you must consult your teacher.

Assessing your report

Your independent investigation report is the Non-exam Assessment (NEA) of the A-level. The report will be assessed internally by your Centre using criteria and a marking grid provided by OCR. A process of external moderation will check that marking standards are applied equally across the Centres.

Your report will be marked out of 60, which represents 20% of the A-level qualification. The marks for each section are:

- Planning, purpose and introduction — 8 marks
- Data, information collection methods and sampling framework — 17 marks
- Data presentation techniques — 9 marks
- Data analysis and explanation — 14 marks
- Conclusion and investigation evaluation — 12 marks
- Overall quality and communication of written work — 10 marks

The marking criteria for the top level are included at the end of each of the main sections in this guide. Four of the sections have three mark levels, the remaining two have four because more marks are allocated to these.

The assessor will mark your report on a 'best-fit' principle. A report does not have to meet all of the requirements of a level descriptor for the mark to be awarded at that level. How well the report meets the requirements of a particular level descriptor will determine which of the range of marks available will be awarded.

Exam tip

As you plan, execute and then complete each section of your investigation, use the marking criteria as a checklist against which you can assess your work.

Planning, purpose and introduction

Lack of clarity in the aims and objectives of your investigation will result in a muddled and often superficial final report. Start investing time and effort from the very beginning.

A successful investigation is often one that is SMART.

S **Simple:** do not choose multiple tasks — a single question or hypothesis avoids muddle.

M **Measurable:** can it be measured in one way or another? This applies to a wide range of data not just numerical data.

A **Achievable:** can you carry out the investigation given the location(s) and resources available?

R **Realistic:** avoid speculation and what ifs. Choose something that can be investigated directly and now.

T **Timed:** you should be able to complete the investigation within the time available to you.

Where do I start?

The key aspect of your individual investigation is that the topic should be one you are interested in. If you are to sustain the level of commitment required to see this project through from start to finish, then enthusiasm as well as hard work is needed.

Choose an area of the specification content that you find particularly interesting. This could come from:

- Component 01 Physical systems
  - Landscape systems: **either** coastal **or** glaciated **or** dryland landscapes
  - Earth's life support systems

- Component 02 Human interactions
  - Changing spaces; Making places
  - Global connections: **either** Global systems (Trade in the contemporary world or Global migration) **or** Global governance (Human rights or Powers and borders)
- Component 03 Geographical debates
  - **Either** Climate change **or** Disease dilemmas **or** Exploring oceans **or** Future of food **or** Hazardous Earth

Once you have chosen a broad subject area you can start to plan your investigation. All too often the approach adopted is a rigid, linear one such as that shown in Figure 1.

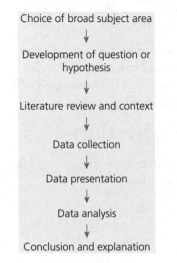

Choice of broad subject area
↓
Development of question or hypothesis
↓
Literature review and context
↓
Data collection
↓
Data presentation
↓
Data analysis
↓
Conclusion and explanation

Figure 1 The linear approach to geographical investigation

Such an approach can result in an unconvincing investigation, as it fails to take account of various things that almost inevitably crop up during the project. For example, during data collection you may discover that you are not getting the data you thought it would be possible to obtain. You might, therefore, need to go back and revise the question. In reality, you must be ready to reflect on the progress of your investigation at each stage and be willing to go back and revisit a previous stage if necessary (Figure 2).

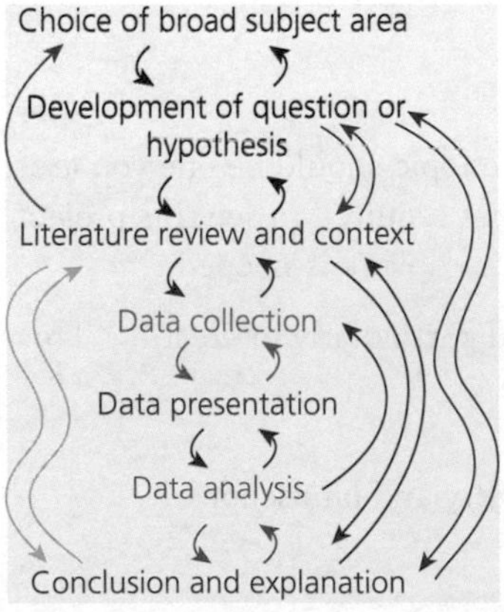

Figure 2 The reflective approach to geographical investigation

Exam tip

Keep a record of any changes you make as they occur. This will allow you to write convincingly when evaluating matters such as data collection methods or data analysis.

Quantitative and qualitative investigations

Some geographical topics are best investigated through a **quantitative approach**. This applies to both physical and human topics, such as changes in sediment size along a beach or variations in demographic characteristics in a rural area.

In contrast, a **qualitative approach** can be more suited to other topics, for example asking people what their feelings are towards a rebranding project or what their views are of local health services.

It is possible that a particular topic is best investigated by using a combination of quantitative and qualitative methods. Qualitative material such as interviews about the use of a local park may need the support of some numerical data about the demographic characteristics of the people in the location.

A **quantitative approach** is based on numerical data and often includes statistical testing and/or mathematical modelling.

A **qualitative approach** gathers information about how people experience and view the world, either where they live or further away.

Is my investigation feasible?

It may not be possible to carry out some investigations because of:

- practical problems: for example, difficulty of obtaining some data, for example rates of cliff erosion or weekly income levels in an area
- inaccessibility: for example, private property
- no secondary data: for example, lack of data from the past, such as journey-to-work flows from ten years ago

Walking your local area

It is essential to visit the location(s) that you are considering for your investigation before proceeding too far through the planning process. Observation in the field will suggest ideas to be developed as questions or hypotheses but will also act as a check on the feasibility of the project. As you walk the location(s) note anything of interest, particularly where a change occurs — for example, where the type of sediment or vegetation changes along a beach **transect**. When walking round a settlement, note where and how building style or land use changes.

A **transect** is a line or path across a feature along which observations are made or measurements taken.

It is useful to observe your location(s) from contrasting viewpoints, such as different weather, light conditions, day of the week or time of day. Changed conditions may suggest new ideas for investigation.

The reasons for the location of certain land uses, such as types of retailing or leisure facilities, may not be readily apparent on a Sunday afternoon. The users of a local park or town square may show significant variations according the day of the week or time of day.

Exam tip

During your walk(s) take photographs and make notes, preferably on a map, of what you observe. These will help you in devising a question and a strategy for data collection.

Some processes in physical geography are spasmodic or seasonal. Landforms can change dramatically over short periods of time. For example, a high-energy storm can significantly alter the profile of a beach. Seasonal changes can, to some extent, be predicted, such as how a woodland alters month by month.

Care must also be taken if the boundaries of the areas being investigated have changed. For example, in 2001 some spatial units used by the Census changed. The enumeration districts for small areas were replaced by units called Super Output

Areas. This can mean that investigations involving changes through time using data collected in censuses may not be successful as you are not comparing like with like.

Why is scale an important consideration?

Choosing an inappropriate **scale** is likely to compromise your investigation in terms of the accuracy and reliability of data.

Scale refers to the size or extent of the area of investigation.

For example, to research changes along a transect, such as along a beach or road, or to compare the demographic characteristics of two or more areas, the scale of the transect or areas must allow differences to be evident. Too short a transect or too small an area will make it difficult to see meaningful variations.

If you choose too large a scale, your investigation is likely to become unfeasible, often in terms of data collection. You could also miss crucial differences amongst data and so come to conclusions that do not reflect the actual geographical pattern.

What is an appropriately defined aim, question or hypothesis?

There are two types of **hypothesis** in geographical investigation:

1 focused on spatial (geographical) differences
2 focused on relationships between variables

A **hypothesis** is a set of beliefs or assumptions about the structure of the world or the way it operates.

A hypothesis should be able to be tested. For example, for a sand dune system the hypothesis 'Depth of soil increases as distance from the sea increases', and in a rural setting the hypothesis 'The larger the population of a settlement, the more services it will possess' can be tested.

A clearly defined hypothesis will show the direction of the differences or relationships being investigated and thus will steer a project in more productive directions. For example, 'Shingle beaches have steeper gradients than sand beaches' or 'The age of buildings decreases with increasing distance from the centre of a settlement'.

The null hypothesis

Scientific methodology encourages forming two hypotheses, one positive, one negative. The negative or null hypothesis (written as H_0) states that there is no relationship between variables. It is this statement that is tested. While it may seem rather strange to set up an investigation like this, the key aim of the null hypothesis is to give research a more objective approach. Simply having a positive statement — that there is a relationship between variables — has been shown to influence a researcher's bias so that people see a relationship just because they want to.

If the analysis shows that H_0 cannot be accepted, then the positive or alternative hypothesis, H_1 is accepted and H_0 rejected.

Often, asking a question can give a clear aim to an investigation (Figure 3).

- What is there and why?
  - What are the retailing characteristics of a place and why?
  - What is the micro-climate of a small wood and why?
- What are the differences and why?
  - What are the differences in vegetation across a salt marsh and why?
  - What are the demographic differences between two local places and why?
- What is/are the association(s) between two variables?
  - How do soil depth and slope angle vary across a glaciated valley?
  - How do socio-economic characteristics and healthcare vary within a settlement?
- How are variables associated and in what ways are they related?
  - How are sediment shape and size and slope angle related?
  - How are slope angle and land use related across a farm?

Figure 3 Types of questions

Questions/hypotheses based on theories, ideas, concepts or models

There are various **theories** (often including **laws**) in geography which could be used to generate a hypothesis or question.

Central place theory, for example, could be used as a base from which to derive hypotheses or questions about locations within settlement hierarchies with the aim of investigating different peoples' perceptions of place.

Models have been used in geography as ways of trying to organise spatial patterns and processes.

The gravity model is an example of a model that can be investigated locally. It was devised to represent a wide range of flow patterns, such as movements of people for retailing or employment, telephone and internet traffic, or flow of goods. The population figures for settlements could be used as the 'mass' in the model to predict the volume of flow between places and this compared with the results of an investigation.

There are idealised or model representations of how vegetation changes across both sand dune and salt marsh ecosystems. An investigation can compare a local ecosystem with the model one and explore reasons why differences exist between the two.

Many **theories** are sets of statements or principles that have generally been tested and shown to be true. Others offer ways of understanding aspects of the world and are designed to help understand patterns and processes without necessarily having the ability to predict.

Laws are statements that are universally true, such as 'energy can be neither created nor destroyed but can change forms and flow from one place to another'.

A **model** is a simplification of reality designed to highlight the essential features of something. Some models are timeless while others reflect when and where they were first developed.

Investigations focused on people and the ways in which they interact with other people and places can benefit from considering ideas about such aspects as perceptions, the influence of gender or the role that psychology can have. Similar considerations can help guide qualitative investigations, such as an analysis of informal representations of places, for example prose, poetry, art or music.

Literature review

Whatever topic you choose to investigate, there will be a range of literature that is relevant. Textbooks and articles from magazines such as *Geography Review* offer explicitly geographical sources. However, you should cast your net more widely to include publications primarily aimed at other subjects, such as *Economic Review*, *Biological Sciences Review* and *The Economist*. Make sure you record exactly which pages you have made notes from and, if you copy a diagram or quote some text, this must be acknowledged in your report.

Exam tip

Your school/college library should have access to a wide variety of literature, including past copies of journals either in hard copy or digitally.

The internet offers a vast quantity of material but a critical issue is its origin. Be wary of what appears when first searching and be quick to consult your teachers. Your school or college should have staff able to help you use the internet in ways that lead to reliable, accurate and trustworthy materials. Sites with the extension .ac.uk are from British universities which have a very high reputation. UK government sites, national and local (.gov), also contain valuable material. The websites of two organisations, the Royal Geographical Society (www.rgs.org) and the Geographical Association (www.geography.org.uk), have areas designed to help students.

Exam tip

Keep a record of the date when you accessed online material. Because material can be removed from websites it is best that you either make notes when you first read the material or print off a copy.

It is very important, especially for websites, to keep records of which sites and what you have used from them so that you can offer clear evidence of the context in which you have set your investigation.

Local newspapers continue to offer a wide range of source material. It might be that you can interview a journalist who could then offer further contacts, for example when investigating a rebranding scheme.

Any literature, including online material, must be listed in a bibliography. Your teacher will offer you guidance as to which referencing system to use.

Locating your investigation

You must clearly identify where your investigation is based. This can be achieved by using an extract from an Ordnance Survey map. Your school or college may have access to various forms of digitised maps which can be manipulated to generate a map specific to your own investigation. Digitised maps such as Bing and Google can also be suitable sources of geo-located material.

Exam tip

Make sure you are confident in reading maps at different scales. In particular, both for an A-level investigation and geographical skills at A-level and AS, OS maps at 1:50,000 and 1:25,000 should be familiar to you.

The use of a global positioning system (GPS) can be very helpful in the field. Many mobile phones have the potential to show your location to a very precise degree. This can help accurately geo-locate where you collect data from.

It is vital that you do not simply download or copy a map. You should annotate it with information relevant to your investigation such as:

- exact locations where data were collected
- justifications as to why these particular data collection sites were chosen
- comments about why any potential sites were not chosen

It is increasingly possible to use aerial or surface-based photography. These types of images can be valuable in highlighting advantages and disadvantages of locations for data collection.

The scale of the locational material should reflect the scale of your investigation. Maps of the whole country showing where your investigation was carried out are rarely relevant to the investigation. Some projects, however, need a wider context so that factors such as prevailing wind and wave direction can be highlighted. Journey-to-work flows need maps appropriate to the distances travelled. By contrast, an investigation into the sense of place given to a city centre through its architecture requires locational material focused on the precise survey area.

Whatever your investigation, it is likely that it will require maps at different scales so that you can highlight contrasting but relevant information.

Exam tip

A hand-drawn locational map can be very effective as it can focus on locations and features directly relevant to your investigation. Simply work out an approximate scale when drawing this and include it on your map.

Managing your time

Carrying out an investigation is quite different to most if not all of the other components making up an A-level course. While some aspects will be carried out with teachers and fellow students, typically much of the work is tackled individually. Although your school or college is likely to have a broad timetable for the investigation, you are responsible for seeing the project through to a successful conclusion. You need to plan and organise your own timetable.

Planning your timetable

Starting with the submission date, work backwards to when the investigation is introduced. It may seem that there is plenty of time, but it will pass by very quickly. Rushing at any stage is unlikely to gain you a high mark. A week-by-week timetable will give you short-term targets and help you develop and sustain a strong forward momentum through the investigation.

The various stages do not necessarily require the same amount of time allocated to them. However, do not underestimate the demands of what may seem to be straightforward tasks. Drawing some graphs can be fraught with complications, such as choosing inappropriate scales or it being impossible to label individual points clearly. You are then faced with having to start again.

Data collection can often be a frustrating experience because of practical issues. Adverse weather, not enough people to interview or equipment not working will mean having to come back another day.

Before you start adding the various stages of the investigation to your timetable, add in all the commitments from other areas of your life. Events with family and friends, sport, drama, music or employment will need to be taken account of.

Exam tip

Make sure you build 'slippage' time into your timetable to deal with the unexpected such as illness or computer issues.

Above all, be organised so that you can make the most of all the hours you will be putting in to your investigation.

Minimising potential risks in fieldwork

Risks cannot be entirely eliminated but they can be minimised by taking sensible precautions.

1 Identify potential hazards.
2 Assess the level of risk for each potential hazard.
3 Devise a strategy to minimise risks.

Your school or college will probably have a risk assessment form which will help you in managing fieldwork risks. Consulting with your teachers will also help you to assess risks and plan appropriately.

The nature of risk varies according to the location(s) fieldwork is conducted in. There are, however, some general principles that can be helpful when planning data collection.

- Give your teacher or other responsible adult precise details of your travel to and from fieldwork sites and when and where you will be collecting data.
- Consult the weather forecast on the day of the fieldwork. Tide times are available well in advance and it is best to check with someone who knows the local conditions, such as at a local sailing club.
- Have a fully charged mobile phone with you and leave it switched on. Make sure you have appropriate telephone numbers and that your teacher or responsible adult has yours. When you first walk a potential area for investigation, check the quality of your mobile connection.
- Take appropriate clothing, including footwear. Layers of clothes, including gloves and a hat, are important as these can be added or removed as the weather changes. A waterproof outer layer is essential. Think about the environment you will be in and choose sensible footwear. Boots that offer good grip and ankle support are required where the ground is rough and/or steep. Wellington boots are an option when investigating coastal or river environments. Hard hats are required in locations where there are overhead hazards such as cliffs.
- If working in a remote location carry a torch, survival bag, whistle and emergency rations.
- Carry a small first aid kit. Include sun cream and apply this before setting out if the day has little cloud.
- Work in groups, ideally of at least three people. If there is an accident, one person can stay with the injured while another goes for help. Remember that mobile phone networks do not operate in some locations.

Exam tip

It is important that you are advised in your risk assessment by your school or college.

Risks to data collection

Risk assessments should also take into account risks to data collection. For example, what if an item of equipment breaks or stops working the morning before data collection is complete? If it is pouring with rain on the day a questionnaire survey is planned, people will be very reluctant to stop to take part. Emergency roadworks or a traffic accident will alter traffic flows. Always have a back-up plan so that data collection is not compromised.

Risks to the reliability and accuracy of your data could undermine the investigation. This is why time spent walking potential locations and conducting pilot surveys are important, as these help identify potential risks to data collection.

Some time spent considering risks and planning appropriately will minimise risks both to yourself and your investigation.

Marking criteria

Section 1: Planning, purpose and introduction	
Level 3 (6–8 marks)	■ There is a clear, well focused plan, appropriately designed to include aims or questions or hypotheses linked to the geographic purpose of the investigation. ■ The plan is based on an individual geographical topic or issue, which is accurately and appropriately defined and within a research framework. ■ There is a justification for the investigation provided in the introduction and valid contextualisation of fieldwork and research. ■ The location is precise and geo-located, using geo-spatial techniques at appropriately different scales. ■ There is clear evidence of valid and individual literature research that defines and contextualises the investigation through an appropriate combination of wider geographical links, comparisons, models and theory.

Data, information collection methods and sampling framework

Once the hypothesis/question is established, you then need to decide what information and data will be required in order to test/answer it. You can then work out how to collect your information and data.

Two critical questions apply to data collection.

1 Are my data **reliable**?

2 Are my data **accurate**?

Reliability refers to consistency. If data collection was to be repeated, would the results be reproduced? The most serious threat to reliability is bias, so having a clear and rigorous framework for your data collection will help reduce this risk. For example, test and calibrate equipment such as a thermometer to make sure it is working before going out into the field. The issues of sample size and method are also relevant.

Accuracy refers to whether the values recorded for your data are correct. Have you used the instruments in ways that they are designed for? Have you counted the traffic flow accurately? Being consistent in how you measure is important for accuracy. How far into the ground do you push the ranging pole? For pedestrian flows, do you count children in pushchairs?

Knowledge check 1

What is meant by primary and secondary data?

Knowledge check 2

What is meant by quantitative and qualitative data?

You must be unambiguous when you record data. '30 centimetres' could be recorded as '0.3' but can you be sure what you measured several days or weeks later? This is especially important if collecting data in groups. Make sure everyone writes clearly or uses electronic recording carefully and consistently. It is a good idea to take turns in carrying out the various tasks involved in data collection. You will then be in a stronger position to assess the strengths and weaknesses of data collection methods such as the reliability and accuracy of equipment.

Ethical and socio-political implications of data collection

Your investigation is not set in a vacuum separate from the real world. It is influenced by ethics, moral principles and rules of conduct governing behaviour. You must be aware of these as you plan and execute your work.

When using techniques such as interviews and questionnaires, it is important that you appreciate the rights of respondents. They must give 'informed consent' before taking part. You must make them aware of the type of information being collected, how it will be used, where it will be published, that their anonymity will be protected and what will happen to their data once the study is complete.

Consider particularly carefully what personal information is required in your investigation. Names, addresses, postcodes, occupations, age, income, for example, can be sensitive data items.

If you are surveying residential areas remember that these are peoples' homes. Be thoughtful in the way you conduct your data collection, for example when taking photographs.

Some topics have socio-political implications that make some data too sensitive to ask about. Factors such as race, sexuality, religion, political affiliation must be treated with respect and caution.

There are also ethical matters to consider in fieldwork that is not focused on people. You must think about your environmental impact. Trampling vegetation or disturbing features such as river banks and scree slopes must be considered and plans made to avoid or minimise your impact.

Exam tip

An annotated photograph can illustrate the issues you might have faced in minimising your impact on a location when collecting data.

Piloting your investigation

Time and effort spent in advance of the main data/information collection will repay itself many times over. Trying out any equipment you need to use beforehand will soon highlight any practical issues. Conducting a pilot survey of questionnaires or traffic counts, for example, allows you to evaluate your techniques. Which questions were successful in generating material and which were not? Was the form you were using to collect data effective in the field?

A pilot may also highlight risk factors that need to be assessed and the steps required to eliminate or minimise them.

Qualitative approaches

These are most appropriate for people-centred investigations. They rely on observation, interviews and textual analysis. Investigating informal representations of places would be well served by the use of qualitative methods.

Observation

The observation of people and places can involve several techniques, such as:

- notes made in the field
- field sketches
- photographs you take
- audio-visual recordings

These observations are known as **naturalistic**.

Naturalistic observation notes people's behaviour in a real-world setting such as a park or a retail centre.

Observation has the advantage of simply letting people go about their lives without a researcher's involvement which may alter their behaviour. However, you need to be aware of disadvantages:

- it can only be used in small-scale investigations — there is a limit to how much and how far you can observe at any one time
- it is difficult to repeat the observation in exactly the same way at another time or in another place
- the bias of the observer

Interviews

As part of a qualitative investigation, interviews are likely to be informal. Conversations and discussions can be held with individuals and groups which are probably best recorded verbatim. These can be transcribed later, perhaps using voice recognition software, and then analysed.

You do not need to involve objective statistical sampling in deciding who to interview. It is likely that the investigation will be into a specific local issue, such as the redevelopment of part of a settlement or the closure of a health facility. You need to interview people who are directly involved, such as local community or special interest groups. By interviewing these people, you will research their motives, behaviours, attitudes and perceptions.

Exam tip

Transcribing and/or analysing interviews is best done soon after conducting the interviews. If you leave it too long, you may not be able to recall points made in the interview that are unclear.

Textual analysis

Textual analysis is a method of describing and interpreting the characteristics of a recorded or visual message. The purpose of such an approach is to describe the content, structure and functions of the messages contained in texts, including visual material of all types such as paintings, newspaper photographs or graffiti. It can also include analysing material originating from social media such as blogs.

Various questions can be asked when analysing images such as photographs taken by others or by yourself. For example:

- What is the setting of the photograph?
- What is the likely time of year and day?

- What is the subject of the photograph?
- Does it include people, animals, buildings and/or scenery?
- What is the main activity seen in the photograph?
- Does there seem to be a theme to a set of photographs?

Knowledge check 3

What are the three principal methods of collecting data in a qualitative approach?

Coding qualitative material

The diverse nature of a qualitative approach requires rigour in organising the material collected. Coding is central to making the most of qualitative material.

Coding is the process of sifting through your material for themes, ideas and categories. You would decide on a system that identifies and classifies the particular areas you are researching. For example, if your focus was the redevelopment of some aspect of your local area you might want to identify the different players/stakeholders involved and mark all your data with appropriate codes, such as meaningful words, letters, abbreviations or symbols. Positive and negative attitudes towards the redevelopment could be coded. It is then possible to retrieve data quickly for further analysis at a later stage. Coding can also help identify patterns in the material.

As you look through your qualitative material it can be helpful to ask questions such as:

- What is going on?
- What is the main/secondary focus of the image?
- What are people doing?
- What advantages/disadvantages are people identifying?
- What are people taking for granted?

It is likely that you will start with a set of codes you have chosen from your background reading and ideas that emerged during the planning and conducting of the data collection. However, always be open to something emerging from the data as you look through them and be ready to devise a new code.

Knowledge check 4

Why is coding important to the success of a qualitative approach?

It is important to write notes throughout the coding process. Definitions of codes, issues you found when applying codes or ideas about people's attitudes that struck you must be noted as you are unlikely to remember them later!

Throughout the coding process, be aware that it can be strongly influenced by the bias of the researcher, which is you. Coding is also a time-consuming task and may require you to revise your ideas and start again — an example of a reflective approach.

Quantitative approaches

Many investigations require data collection through the measurement of variables.

A **sample** is a subset of the population and acts as a substitute for the entire population.

Population, in the context of investigation, refers to all the data that could be measured.

Sampling methods

Most A-level investigations are based on **samples** of a **population**.

It is often neither necessary nor practicable to count or measure the entire population, for example pebbles on a beach, scree fragments on the side of a glaciated valley, slope angles across a farm or households in a settlement. The total population size or locations may not be known. It might be impossible to gain access to the entire population.

Various ways of sampling have been devised with the aim of delivering a sample that is representative of the entire population. You can then be confident in your conclusions. To be effective, sampling needs to:

- produce a large enough sample to give conclusive results in terms of statistical significance
- be unbiased, i.e. every individual member of the population has an equal chance of being measured. This will give estimates of the characteristics of the population (e.g. mean and standard deviation) that are neither consistently greater nor smaller than the true values
- be precise, so that it offers an accurate estimate of the characteristics of the population
- be collected easily with the minimum of resources

Sample size

Sample size is an important consideration in research. It is a widely held view that the bigger the sample the better the study. This is not necessarily true. In general your investigation will be more successful if you improve the accuracy of data collection, for example practise using equipment or conducting interviews.

In qualitative investigations, the key aspect is the richness of the source of data, be that a person, a visual image or some text. You can be pragmatic in deciding your sample size, considering factors such as time taken to conduct, transcribe and code interviews. The number of stakeholders (players) is another factor to think about.

Exam tip

Know the minimum amount of data required by any statistical tests you might use in analysis. Make sure when planning your sampling method that you collect over that minimum.

In quantitative investigations, there are various statistical methods for determining the size of a sample. However, these are more appropriate at university level and beyond, such as in medical research. For your A-level investigation you need to think about the expected patterns you observed when you 'walked the site(s)' and then reflect these variations in your sampling. This might include, for example, how many transects you have and how frequently you measure a variable along a transect.

In an A-level investigation, it is necessary to be pragmatic about sample size. Collecting infiltration rates or noise levels at just two sites on one day at mid-day, clearly is inadequate. You therefore need to decide what the influences are on your potential data/information and how these might vary both through time and space. Some rational decisions can then be taken about sample size, also taking into account the resources required to collect the sample material, such as time and equipment.

Bear in mind that most statistical analysis requires a minimum number of readings for it to be valid. You are on safer ground if your sample size is comfortably over that minimum. It is better for the reliability of your investigation that a larger rather than a smaller sample is collected.

Exactly which sampling method you use will depend on the data required for your particular investigation. Sampling methods can be divided into two categories, spatial and non-spatial (Figure 4).

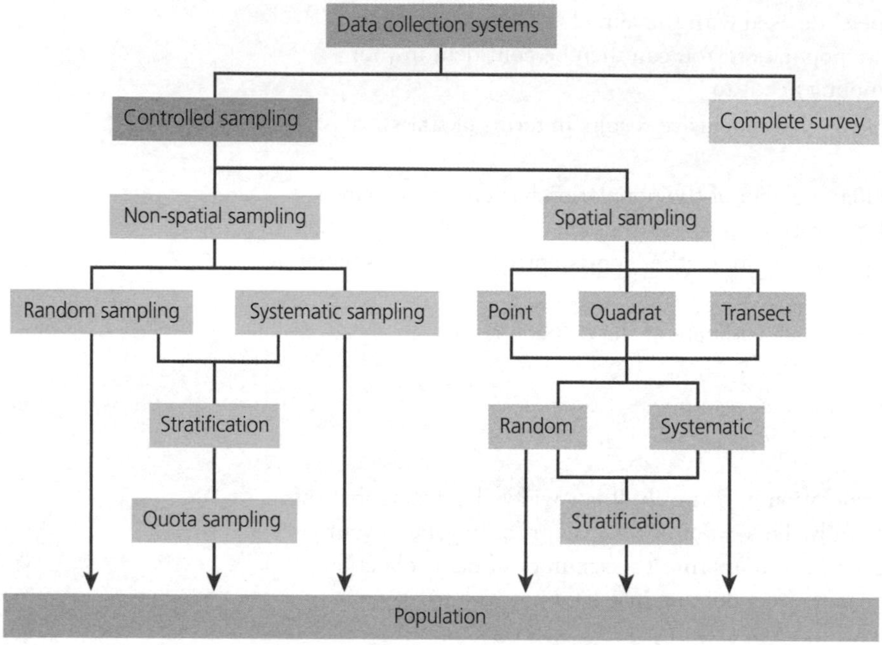

Figure 4 Sampling methods

Non-spatial sampling

There are three types of non-spatial sampling:

1 random sampling
2 systematic sampling
3 stratified sampling

Random sampling

The assumption of random sampling is that every item in the population has an equal chance of being selected — rather like pulling names out of a hat. Obtaining a random sample is usually achieved by using a random-number table or from a calculator. This avoids bias as any personal preferences of the researcher are excluded. It is possible, however, that a random sample may not represent the population as a whole. The sample may miss out some critical aspects of the population. For example, in an investigation into recreational activities of different households, a random sample may fail to select households without access to a car. Using random numbers can also be a time-consuming process as not every member of the population selected can be used in the survey. For example, you may not be able to get to every household whose address is chosen.

Systematic sampling

This approach chooses individual items at regular intervals, such as every fifth person, five metres or five minutes. It is more straightforward than random sampling and can allow you to cover the population effectively. However, it is more biased as not all members of a population have the same chance of being chosen. This can lead to a misrepresentation of the geography being studied, such as an inaccurate pattern emerging.

Stratified sampling

It may be that you are aware that significant aspects of the geography of your study area exist, such as different soil types or different characteristics of people (age, gender, ethnicity, employment and so on). When the extent or size of these subsets are known, stratified sampling allows your investigation to be proportional and representative of the whole population. Conducting a **pilot survey** can reveal such differences.

A **pilot survey** is conducted before the main research takes place. It uses a small sample and can test the effectiveness of your data collection techniques.

The major disadvantage is that you need to know the characteristics of your population and be able to classify them into subgroups. Investigations focused on people are able to make use of census data and the categories used in this. Even here there are limitations, for example a census is only held every ten years so soon goes out of date.

Stratified random and systematic sampling

If you know the relative proportions of the groups that make up the overall population of your survey area, then it is possible to apply either random or systematic sampling to each group. For example, you can assign every member of your school/college to a group such as male student, female student, male staff and female staff. Suppose female students comprise 45% of the total population and your sample size is 100. Your data collection must therefore survey 45 female students. You can either select these randomly or systematically.

Quota sampling

Using this approach means that you choose the individual items to survey. You still have to gather the data in proportion to the representation of the subgroups in the population as a whole but do not need to do this randomly or systematically. For example, when investigating attitudes towards the closure of a local library the sampling frame needs to be divided into four groups depending on age — 13 to 18; 19 to 30; 31 to 65; 66 and over. Each age group is to have 25 respondents, giving a total sample of 100. Both genders need to be represented equally in your sample groups and so a quota of ten males and ten females need to be recruited for your investigation. This more practical approach can save time and money but may suffer from bias in which individuals within the subgroup are chosen.

Spatial sampling

For some investigations, the location of the items being surveyed is an essential aspect. Points, lines or areas can be employed in random, systematic or stratified ways.

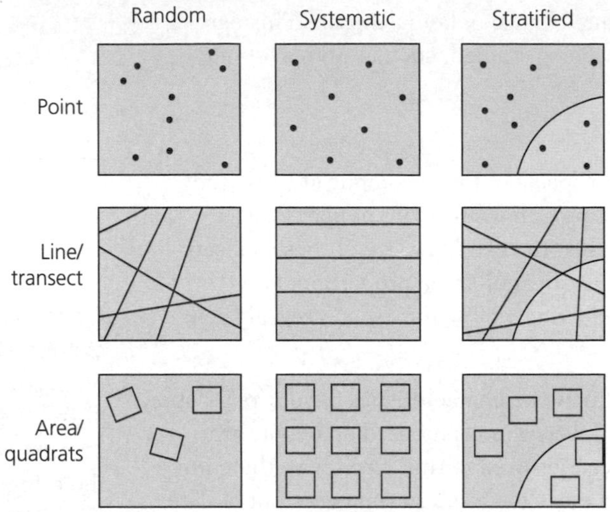

Figure 5 Spatial sampling techniques

Generating random samples can be misleading and time-consuming. It can result in an uneven distribution of samples across the survey area. It can be time-consuming if a large data set is required.

Systematic spatial sampling can be effective, such as equal spacing along a transect. However, deciding on the distance between survey sites needs careful thought. Too close together may make data collection very time-consuming, too far apart and key differences in the pattern can be missed.

As with non-spatial stratified sampling, the spatial equivalent relies on having prior knowledge of the subgroups in the survey area, such as distribution of geology or demographic characteristics.

If your investigation is likely to use a type of isoline map when presenting the data (for example, temperature variations within a wood or distribution of litter in an area), it is important to have sufficient survey points to allow you to draw accurate isolines. Having too few points will mean that you will have to 'guess' the pattern of lines.

Knowledge check 5

What is the difference between non-spatial and spatial sampling?

Spatial sampling and GIS

Geographical Information Systems (GIS) are integrated computer packages for collecting, storing, processing and analysing geographical data that can be plotted on maps. They have become increasingly available and can be an invaluable tool in an investigation. You must, however, spend some time learning the GIS package so that you are confident in your ability to use it appropriately for your investigation. A mobile GIS package such as Survey123 for ArcGIS, could be combined very effectively with a well-designed sampling framework to record a wide variety of data.

Questionnaires

A questionnaire is a set of questions designed to generate data/information from people about themselves and their views of an issue. They are widely used commercially to gather information that might be useful in market research, for example.

Questionnaires are valuable for collecting:

- basic characteristics of a population, e.g. gender, occupation
- patterns of behaviour, e.g. distances travelled, routes taken, frequency of use of a feature
- reasons for behaviour, e.g. preferences for use or not of a facility, preferences for particular routes at different times of the day
- range and strength of feelings towards an issue, e.g. rebranding project, flood prevention scheme

Questionnaires may appear to be a straightforward technique. They are, however, demanding of time and effort to design effectively and you are not likely to get this right first time. Conducting a pilot questionnaire survey (about ten) is vital as a way of finding out which questions generate the material your investigation needs and which do not.

Exam tip

Remember that questionnaires are different to interviews. Questionnaires are a set of pre-determined questions. An interview is a one-to-one conversation of mainly open-ended questions and often involves follow-up questions.

Question types

Most questions are either open or closed.

Table 2 Advantages and disadvantages of open and closed questions

	Open questions	Closed questions
Advantages	■ Flexible, allowing respondents to use their own words ■ Authentic responses, as respondents use their own words ■ May not require large sample if focus is on attitudes/preferences	■ Avoid uncertainties of open-ended questions ■ Take less time to complete ■ Responses lend themselves to statistical analysis
Disadvantages	■ Difficult to classify responses in numeric terms so statistical analysis difficult ■ Questions can be subjective ■ Respondents may not understand them	■ Rigid and restrict respondents ■ Can lead respondents unconsciously to predetermined answers ■ Respondents unable to express opinions

Combining open and closed questions in a questionnaire can generate a great deal of valuable material for presentation and analysis. Figure 6 shows an example of such a questionnaire. It forms part of an investigation into the threat to the identity of a location popular with tourists which is causing concerns for the local community.

Good morning/ afternoon. I am an A-level geography student from x school/college.
I want to conduct a survey of visitors and I would value your opinion.
Would you mind answering some questions? It won't take more than a few minutes.

Use a first sentence as an introduction to approach people. Do not forget to explain why you want their help and why you value it

(1) Have you ever visited this site before? ☐ Yes ☐ No

(2) How did you come to hear about this place? ☐ Advert/flier ☐ Saw on TV ☐ Newspaper ☐ Friend ☐ Already knew ☐ Other

Open with some easy, non-sensitive closed questions to 'warm up' and relax the respondent

(3) How far have you travelled to get here today? ☐ <5miles ☐ 5–20 miles ☐ 20–30 miles ☐ >30 miles

(4) If you did not arrive here by car, how did you travel to the site? (indicate all that might apply)
☐ Bus/coach ☐ Walking ☐ Bicycle
☐ Motorbike ☐ Train ☐ Other

Always include 'other' categories for multiple- answer questions

(5) What are your reasons for visiting? (indicate all that might apply)
☐ Dog walking ☐ Visiting friends ☐ Running/jogging
☐ Sightseeing (little walking) ☐ Walking > 2 miles ☐ Picnicking
☐ Mountainbiking ☐ To use facilities ☐ Taking a rest
☐ Education/scientific ☐ For the children ☐ Nostalgic ☐ Other (specfiy)

(6) If you had more time here, could you rank what your main preference would be from the list on the previous questions?
☐ Dog walking ☐ Visiting friends ☐ Running/jogging
☐ Sightseeing ☐ Walking ☐ Picnicking
☐ Mountainbiking ☐ Facilities ☐ Take a rest
☐ Education ☐ Children ☐ Nostalgic ☐ Other

The questionnaire shows progression — starting with simple closed questions and building towards more open and structured responses. Question 6 is an example of a ranked relative preference

(7) How long do you intend to stay (a) at this site, (b) in the area?
(a) ☐ <1 hour ☐ 1–2 hours ☐ 2–4 hours ☐ >4 hours
(b) ☐ <1 day ☐ 1–2 days ☐ 2–4 days ☐ Longer

(8) Are you aware of any problems with the site? If so what?
[ ]

Question 8 is an example of an open question. Make sure you leave enough room for the response

(9) This is considerd to be an attractive site. How far do you agree? (Circle as appropriate)
Strongly agree | Agree | Haven't considered it | Disagree | Strongly disagree

Question 9 is an example of a 'Likert scale' question — good for attitudes and opinions

(10) Could you give me your postcode? [ ]

The postcode is a very useful piece of informaion with which to work out exact locations

☐ Male ☐ Female
☐ <20 ☐ 21–35 ☐ 36–50 ☐ 51–65 ☐ >65

Avoid asking personal questions directly. You can make an informed judgement yourself

Time and date [ ]
Weather [ ]
Survey no. [ ]

Thank you for your time. All this information will be treated as confidential. Goodbye.

Figure 6 An example of a well-designed questionnaire

People's values, opinions and perceptions can also be recorded using a **Likert scale**.

This assumes that a person's strength of feeling about something can be placed somewhere along a continuum with a neutral point in the middle. The scale is perhaps most effective with five options but can have four choices, forcing the respondent to decide if they lean more towards one end of the scale or the other. A disadvantage of Likert scales is that people may tend towards the centre ground, not wanting to give an extreme reply.

A **Likert scale** uses fixed choice responses designed to measure attitudes or opinions.

Table 3 Examples of Likert scales

Agreement	Frequency
■ Strongly agree ■ Agree ■ Undecided ■ Disagree ■ Strongly disagree	■ Very frequently ■ Frequently ■ Occasionally ■ Rarely ■ Never
Importance	**Likelihood**
■ Very important ■ Important ■ Moderately important ■ Of little importance ■ Unimportant	■ Almost always true ■ Usually true ■ Occasionally true ■ Usually not true ■ Almost never true

The results can be presented using a bar chart and analysed using the median or mode, but not the mean.

Digital and geo-located data

The range of digital data, information stored and used by computers, is vast and increasing. Much of these data have a geographical component — that is any information regarding location. As computer processing power has rapidly improved, recent technologies such as remote sensing (RS) and global positioning systems (GPS) have enabled spatial information in digital forms to be collected. This has led to the development and use of GIS.

Digital data

Mapping is now highly digitised. Part of the Ordnance Survey website (www.ordnancesurvey.co.uk) is dedicated to helping students with mapping. Your school/college may subscribe to the Digimap service, which will help you access OS mapping of Great Britain. This allows the selection and manipulation of information to generate maps specific to your investigation.

National, regional and local government organisations make data available digitally. The Office for National Statistics (ONS) is the principal source of data, much from the census. These data range in scale, allowing you to put together or separate out areas of different population sizes (Figure 7). Access to the statistics can be either as numerical data or as maps.

Exam tip

Make sure you pre-test any techniques collecting data about people's feelings. This will allow you to reflect on how effective your scale has been and to revise it if necessary.

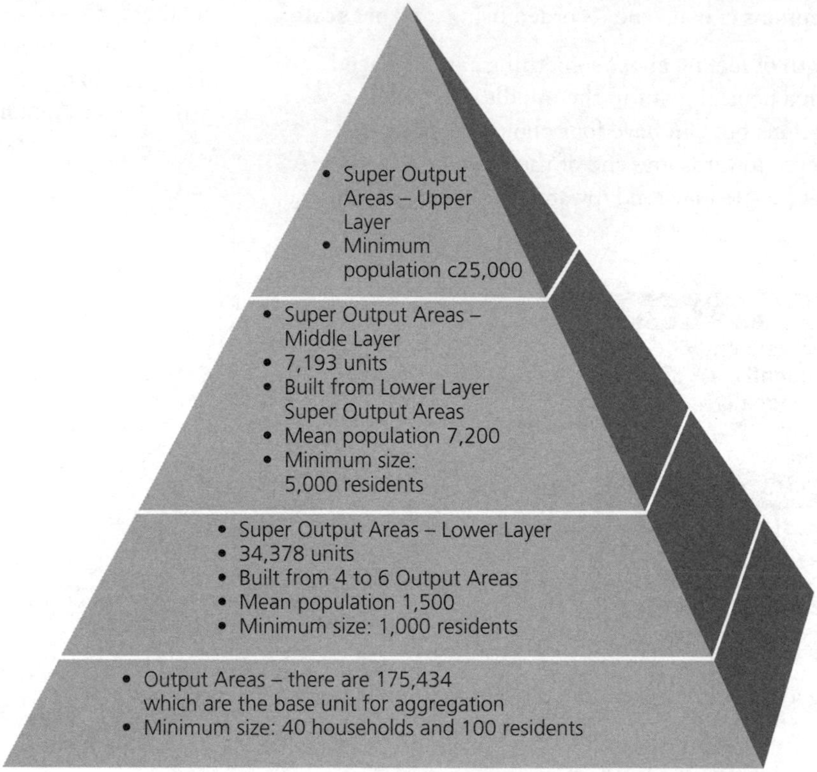

• Super Output Areas – Upper Layer
• Minimum population c25,000
• Super Output Areas – Middle Layer
• 7,193 units
• Built from Lower Layer Super Output Areas
• Mean population 7,200
• Minimum size: 5,000 residents
• Super Output Areas – Lower Layer
• 34,378 units
• Built from 4 to 6 Output Areas
• Mean population 1,500
• Minimum size: 1,000 residents
• Output Areas – there are 175,434 which are the base unit for aggregation
• Minimum size: 40 households and 100 residents

Figure 7 Hierarchy of statistics for the 2011 Census

Big data

'Big data' is something of a vague term because of the different ways it is used. Some people use it to emphasise the enormous scale of data sets that now exist. To others, big data is what might be termed 'found data', the information that is left behind digitally from web searches, electronic payments and mobile phone use, for example. These data result from the ways people's lives have moved to the internet and all these digital exchanges can be recorded and quantified.

Referencing big data has much potential in geographical investigations. It offers huge volumes of data, recorded in real time and usually with locational information. However, most big data sets are owned by large corporations who do not share their material because they use the information for commercial purposes. There are data available such as the famous Google Flu Trends. However, as with Flu Trends, these data may not be up to date and would have to be used as examples of past patterns. Google Flu Trends failed to predict the 2013–14 flu season and is no longer published but past data are available. Big data continues to be researched and like all data has potential and problems.

Crowd-sourced data

The growth in use of the internet and mobile devices has allowed the generation of data based on people's activities and observations. For geographical investigation, these data require a locational reference. For example, the city of Boston,

Massachusetts, has been using a smartphone app, Street Bump, which records when a car hits a pothole in the road. The information is sent to the highways department and a map generated for road repairs.

Cycle share schemes in many cities around the world use online data in real time to indicate the current availability of cycles, which allows people to make informed choices about where to go to find a cycle.

In the UK, the Forestry Commission has a Tree Alert online system for people to report problems regarding the health of trees such as ash dieback disease, including the location.

Twitter can be a source of data from large numbers of people. Because of the nature of social media it is an interesting source of opinions and perceptions of places. Hashtagging (#) a location reveals images and texts that can be analysed.

However, as with all data, care must be taken in analysis. Big and crowd-sourced data must not be assumed to represent all the potential data. There is a degree of sampling bias, as some potential data are missing. For example, the resulting map of potholes in Boston is biased in favour of areas of the city where there are more affluent people who own smartphones and cars.

Knowledge check 6

Why do you need to be particularly careful in using crowd-sourced data?

Geo-located data

Smartphones and tablets

The in-built location and GPS functions, cameras and notes apps in these devices have great potential in providing geo-located data. Google Earth is a source of digital maps, photographs and satellite images, all with locational data.

The British Geological Survey (BGS) offers a free download of its iGeology app. This allows you to consult maps of both solid and superficial geology in the field (www.bgs.ac.uk/igeology/).

Geographical Information Systems (GIS)

GIS is a way of managing the input, processing and output of geographical data. Within GIS there are four main tasks that need to be carried out:

1 input of geographical data, e.g. from paper maps, surveys, satellites
2 storage of geographical data in ways that allow their retrieval, updating and editing
3 manipulation and analysis of geographical data
4 display of geographical data as tables, graphs or maps

Essentially, GIS transforms data into information about spatial patterns. These can then be analysed and conclusions and decisions arrived at. The digital basis of the data means that multiple combinations of data can be generated relatively efficiently. This allows many questions to be asked about data, and the complexity of the real world can begin to be known and understood.

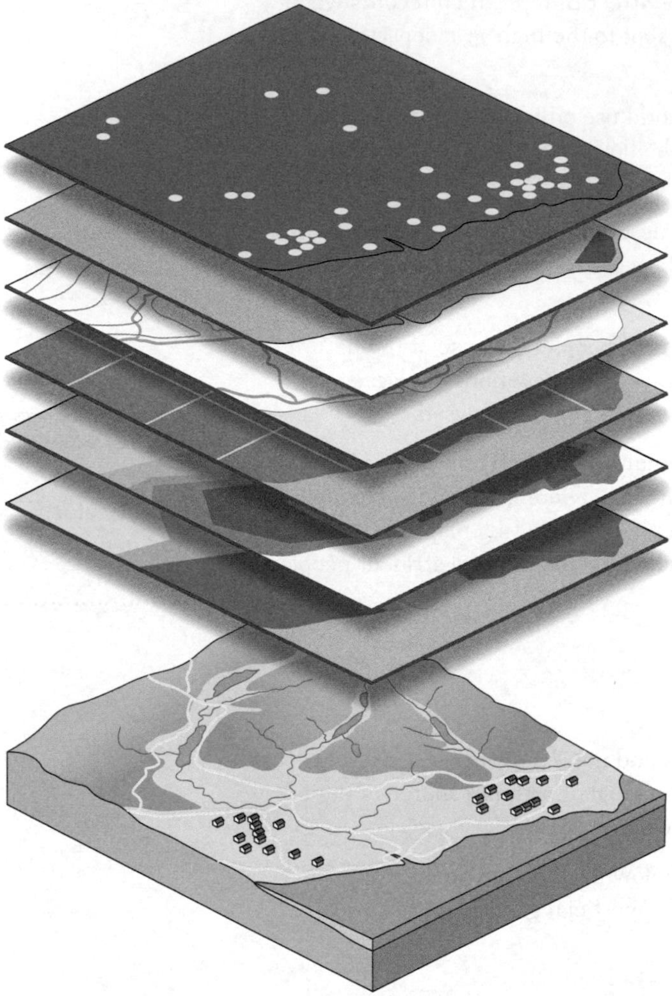

Figure 8 A Geographical Information System

Many websites can be valuable sources of geo-spatial data and are often available with free access. The Environment Agency posts a range of environmental data in map form (www.environment-agency.gov.uk) such as flood risk, coastal erosion, pollution incidents and air pollution. The government site www.magic.gov.uk contains geographical information about the natural environment. It covers rural, urban, coastal and marine environments across Great Britain.

A wide variety of public services provide location maps, such as fire and rescue services, ambulance and health trusts. The police make geo-located crime data available (Figure 9).

The GIS options for an investigation are many and varied. There are a large number of possible combinations of data. It would be too easy to get carried away and produce large numbers of maps containing geo-located data. It is important to keep a sharp focus on the question(s) or hypothesis so that answering or testing that does not become lost in a forest of digital data.

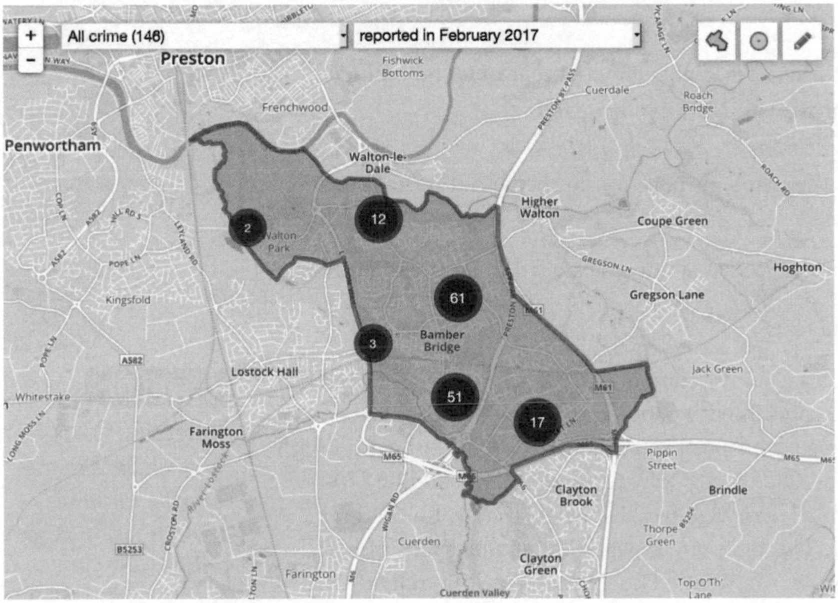

Figure 9 Distribution of crime in a neighbourhood in part of Lancashire

Marking criteria

Section 2: Data, information collection methods and sampling framework	
Level 3 (5–7 marks)	■ There is good knowledge and understanding of a range of data collection methodologies, including suitable quantitative and/or qualitative approaches, which are justified with limitations outlined, appropriate to the investigation and explained in detail. ■ There is clear evidence of personalised methodologies and approaches to observe and record primary data and phenomena in the field and to incorporate secondary data and/or evidence, collected individually or in groups. ■ There is clear evidence of the ability to collect and use digital, geo-located data. ■ The data design framework (sampling, frequency, range and location choice) is appropriate, coherent and justified. ■ Addresses and shows an understanding of the ethical and socio-political dimensions of the methodologies chosen.

Data presentation techniques

Having collected raw data and information, they now need to be processed so that spatial and temporal (if appropriate) patterns, distributions and or trends are revealed. Effective presentation requires the techniques used to be:

- appropriate for the data
- relevant to the question/hypothesis

Maps

Whatever the types of maps used, there are some elements that must be included:

- **scale:** for a sketch map an approximate scale can be given
- **key:** to any symbols or shading
- **title:** ideally this should relate the map's purpose to the theme of the investigation
- **north point**

It is essential to give full details of the source of any maps used in the investigation.

Non-quantitative maps

These present a variety of non-statistical information. Ordnance Survey (OS) publish a wide range of maps at different scales in both paper and digitised versions. The detail they give about relief, drainage, transport, settlements and land use varies with the scale (see pages 507–508 in the *OCR Geography* textbook (978-1-4718-5870-3) published by Hodder Education). There are several specialist maps focused on different aspects of the landscape. The BGS offers paper and digital maps of both solid and drift geology. Road maps are drawn at a large-scale, showing individual buildings in town centres in the UK and some parts of Europe (see page 509 in the *OCR Geography* textbook).

Exam tip

Selecting the appropriate scale of map for your investigation is an important skill.

Quantitative maps

These are drawn using numerical data and are increasingly available digitally, for example in a GIS.

Exam tip

It is vital that the base map on which the data are displayed is appropriate in terms of scale and level of detail. A hand-drawn base map can be the most suitable option.

Dot maps

Each dot represents the same value and is located according to the distribution of the factor being mapped. They can give a very convincing visual impression of a distribution and lend themselves to presenting discontinuous data such as population. To be really effective, however, knowledge about the distribution is required so that spatial patterns and densities reflect the real world. This information may not be known and can be time-consuming to obtain. The value given to each dot should not be too high so that large areas are left blank. In high-density areas dots can begin to merge especially if the value is set too low (see pages 509–510 in the *OCR Geography* textbook).

Choropleth maps

These maps (Figure 10) are widely used to show distributions as they can be very effective at representing overall spatial patterns. They use areal units such as census units and standardised data (percentages, densities, averages, ratios). This means that differences in the data *within* the areal units are not represented and significant elements of a pattern hidden.

Plotting your data on a dispersion diagram can reveal the natural breaks between groups of data. Regular categories can be used but you should check that these do not split a grouping of data. Too few categories and the pattern may not emerge, too many and the map can end up a complicated kaleidoscope with no discernible pattern. The shading should grade from dark (high values) to light (low values) to aid the visual impact of the map (see pages 510–511 in the *OCR Geography* textbook).

Exam tip

It is worth including your dispersion diagram showing the spread of data and where you chose to insert boundaries between categories in your report.

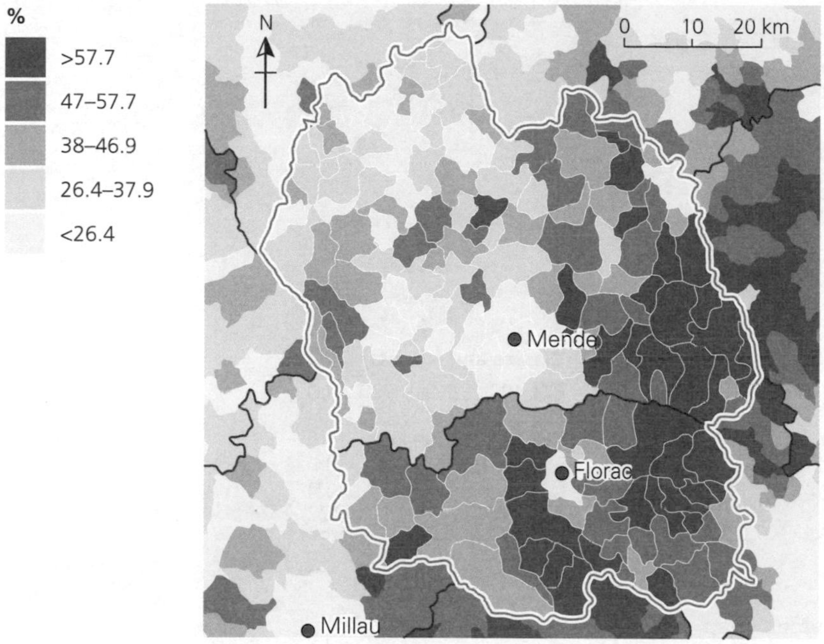

Figure 10 The distribution of second homes in Lozère, Massif Central, France. The data is for communes, a small scale areal unit in the French census

Proportional symbol maps

Whatever the shape of the symbols (circle, square, triangle, bar), if three-dimensional, the areas or volumes are proportional to the value they represent. They show absolute values, not standardised ones as in choropleth maps. This technique can be very effective for an investigation as you can locate the symbols exactly where you collected data, for example along a transect.

The disadvantage of proportional symbol maps is that they can be difficult to read the data values because of their proportional nature. However, this is not a particularly serious disadvantage as the actual statistical data provide that degree of accuracy. It can be difficult to place symbols if they start to overlap one another (see Table 4 and Figure 11).

Table 4 Steps in the construction of proportional symbol maps

Steps	Detail
Source a base map	Source a base map which shows the statistical areas for which data are available.
Choose a symbol	A circle, square, triangle or bar.
Source a scale	The largest symbols should not overlap with the symbols in adjacent areal units by more than 50%. Determine the size of the largest symbol relative to the map (e.g. 10 mm radius for a circle). Calculate the scaling factor for size of symbols based on the square root of the largest value in the data set. Thus, if a circle of 10 mm radius is used to represent a population of 625, the scaling factor is calculated thus: square root of 625 = 25 × 10/25 = 10 (mm).
Placement of symbols	Symbols should be placed centrally within each areal unit.

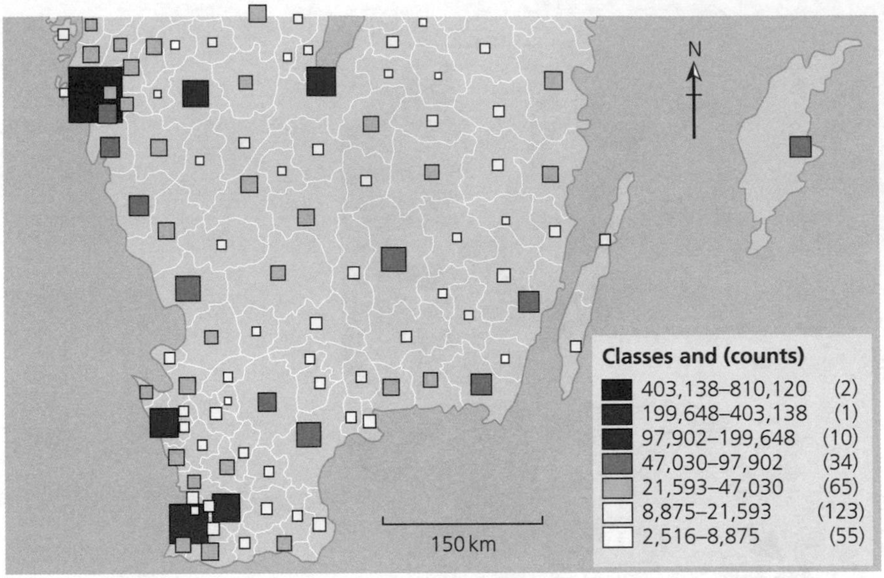

Figure 11 Proportional symbol map: population counts by commune in southern Sweden

Isoline maps

Isolines, sometimes known as isopleths, are lines that join locations of equal value. The most widely known are contour lines on an OS map, indicating height. This is a valuable technique as it conveys an idea of a continuous surface, for example temperature (isotherms) or journey times (isochrones).

Constructing an isoline map (see Table 5 and Figure 12) requires a sufficient number of data collection locations covering the study area. If there are not enough, there is too much guesswork in deciding where the lines should go. The issue of **interpolation** is a serious one and involves thinking about the interval between the values of the lines. It is conventional to use regular intervals. Too few lines and significant elements of the pattern may not show up; too many and you end up with a mass of tightly packed lines that make the map difficult to read.

Interpolation is the process of predicting data values at locations where the actual value is unknown. The assumption is that there is a consistent rate of change between points.

Table 5 Steps in the construction of isoline maps

Step	Detail
Source a base map and locate the data points	Find the location of the data points on the base map, or locate the points from GPS or grid reference data.
Plot the values	Plot the values for each data point on the map.
Decide on the number of isolines and their values	Five or six isolines are usually sufficient. Larger numbers often make construction impracticable. Generally, the fewer the point values on the map, the fewer the isolines. It is conventional to use regular intervals between isolines (e.g. 4 mb intervals between isobars on a pressure chart).
Fit the isolines	Isolines are fitted by interpolation which assumes a constant (linear) rate of change between two points. For example, a 600 mm isohyet between point values of 575 and 625 mm would pass exactly halfway between them.
Number the isolines and add layer colouring	Isolines should be numbered on the map. In order to increase clarity, areas between the isolines may be layer coloured.

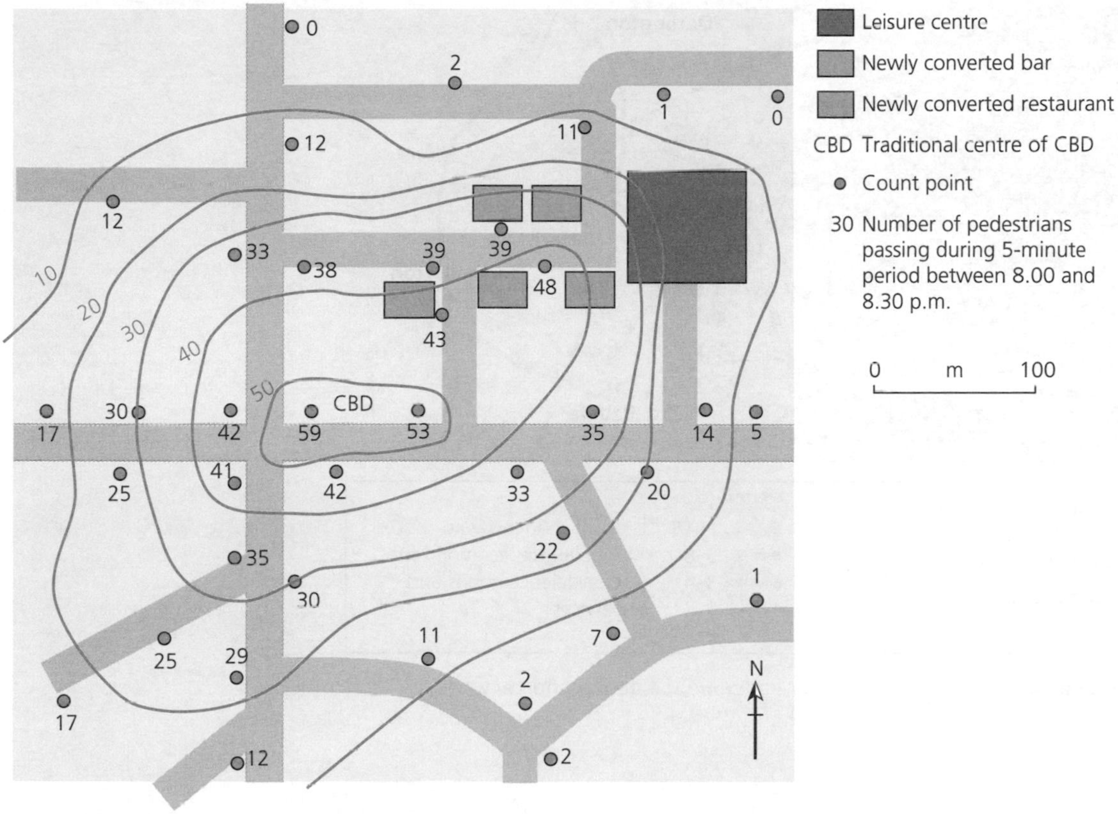

Figure 12 Isopleth map of pedestrian densities in a town centre

Quite often the route of an isoline is unclear from the pattern of point values and it make take several attempts before you have a complete map. Shading between isolines with a sequential colour scheme can help the visual image of an isoline map: darker colours for higher values, lighter colours for lower values (see pages 511–512 in the *OCR Geography* textbook).

Knowledge check 7

What four elements must all your maps include?

Flow maps

These maps (Figure 13) represent movement such as of people, vehicles or information along pathways. They can be plotted as straight-line routes simply highlighting origins and destinations or can follow actual routes. The data can be aggregated so that a particular flow represents all the movement between a source and destination. Alternatively, single lines, each representing the journey of an individual person, can be plotted. This latter style is appropriate when the numbers involved are relatively small.

Visually, flow maps can be very effective but it can be difficult to retrieve statistical data from them. However, their representation of the volume of flow and direction can be helpful in highlighting certain types of spatial patterns and processes.

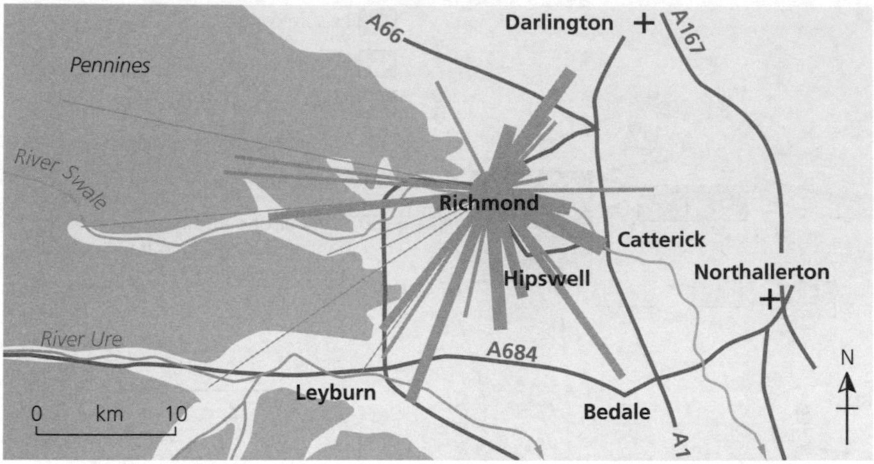

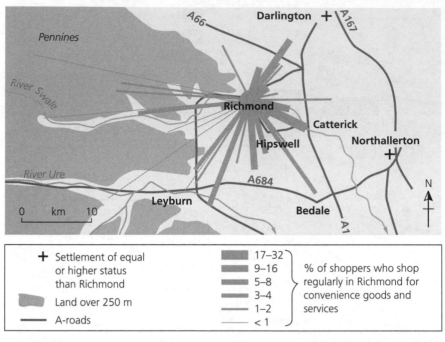

+ Settlement of equal or higher status than Richmond	17–32		% of shoppers who shop regularly in Richmond for convenience goods and services
Land over 250 m	9–16		
A-roads	5–8		
	3–4		
	1–2		
	< 1		

Figure 13 Flows of shoppers for convenience goods and services, Richmond, North Yorkshire

Images

These qualitative forms of data can aid an investigation both in description and analysis. The variety of images available is wide so you must be sure that what you include adds real value to the report and does not take up space simply to 'look pretty'.

Satellite images and aerial photographs can reveal information about land use, settlements, river channels and valleys, and coastlines, for example. **Remote sensing** images offer digital data about a diversity of topics such as land and sea temperatures, vegetation and rainfall intensity. If you think this technology might help your investigation, it is worth contacting your local university to see if there is someone you could talk with.

Remote sensing is a way of collecting and analysing data where the instrument used to collect the data is not in direct contact with the object. Satellites and planes carry instruments such as cameras, scanners and radar of various types.

Photographs

Ground photographs offer both contemporary and historical potential in investigations. Local libraries and museums and published material contain a wealth of historical photographs that could be very useful when investigating changing geography.

Table 6 Making the most of photographs

Taking photographs	■ Make sure images are relevant to the investigation ■ Make sure the features you are investigating are clear and not too far away ■ Include something in the image from which the scale of the feature(s) can be recognised, e.g. metre rule, person, watch (Figures 14 and 15) ■ Make notes: geo-locate, direction you were facing, time of day, weather conditions

Presenting photographs	■ Include figure number, title, location and direction facing ■ **Label** and **annotate** the key features relevant to your investigation (Figure 13) ■ Alternatively, trace off key features and include tracing directly below the photograph. Label and annotate tracing

A **label** is a simple identification of a feature, e.g. floodplain/car park.

An **annotation** gives more detail than a label and includes comments, e.g. ex-village shop converted to residence/rotational slumping leaving exposed scar.

Figure 14 Sediment size, Beer beach, East Devon

Figure 15 Exposed periglacial head deposits, Porlock Bay, Somerset

Labelled and annotated photographs of fieldwork equipment can be useful if they show any issues encountered with data collection, such as obstructions along a transect.

As with all presentational material, photographs should be integrated into the text of the report and not put on separate pages or in an appendix.

Field sketches

These also offer opportunities to describe and analyse data and information (Figure 16). Advantages of field sketches include:

- they encourage you to look more carefully at the area being considered
- they allow you to be selective, emphasising certain features of particular significance and omitting others — a photograph is indiscriminate

Table 7 Making the most of field sketches

Drawing field sketches	■ Before you begin to draw, look at the view carefully — don't be too eager to put pencil to paper ■ Sketch in a few key lines to get some idea of proportion and perspective, e.g. horizon, valley sides, kerb lines ■ Add some idea of scale, vertical and horizontal ■ Add notes: geo-locate, direction you were facing, time of day, weather conditions ■ Field sketches are not works of art! Their purpose is to allow you to offer a representation of some local geography, highlighting the features you consider important to your investigation
Presenting field sketches	■ Add figure number, title, location and direction facing ■ Label and annotate the key features relevant to your investigation

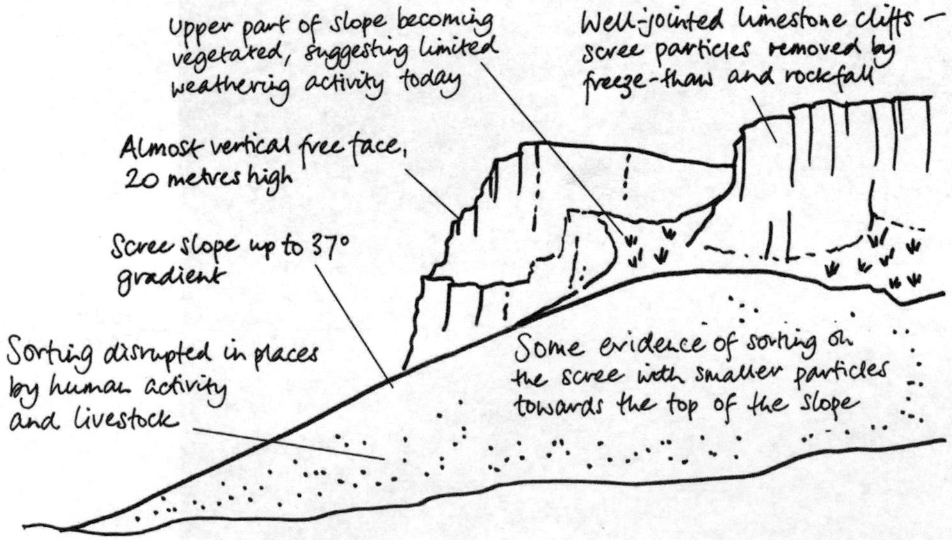

Figure 16 A field sketch of a scree slope at Robin Proctor Scar, Yorkshire Dales

Diagrams and graphs

These are used to simplify complex ideas, patterns and processes. Diagrams can be helpful in highlighting key flows and relationships between and among components in the geography of a location (Figure 17).

Knowledge check 8

What is the difference between a label and an annotation?

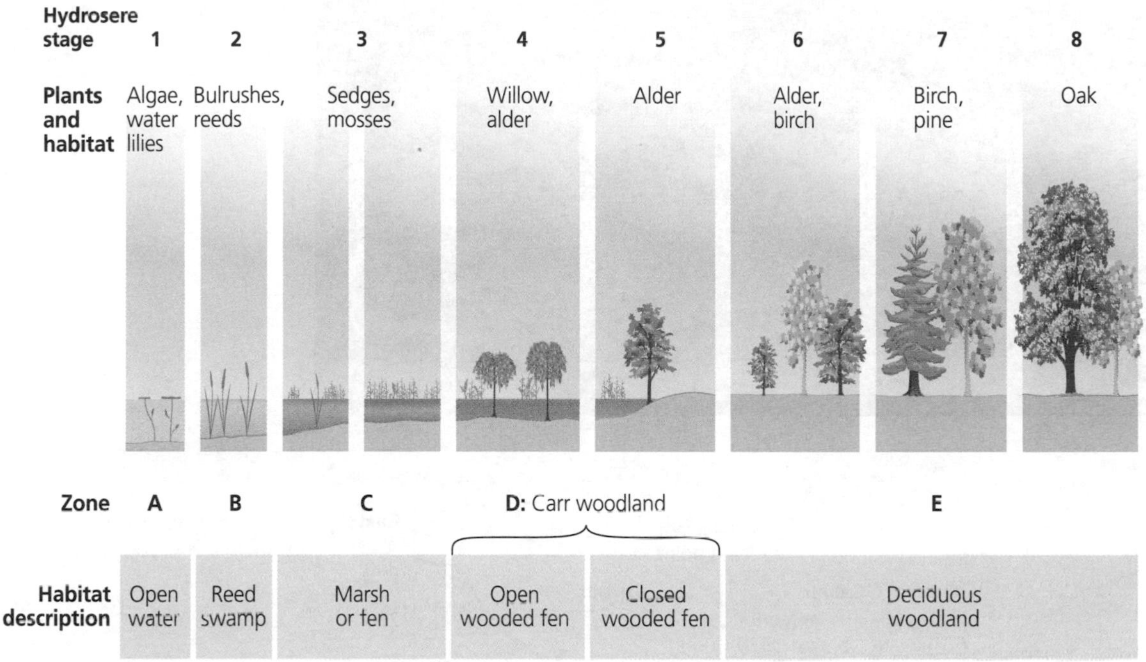

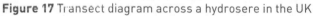

Figure 17 Transect diagram across a hydrosere in the UK

Combining diagrams/graphs with visual information can be very effective (Figure 18).

Kite diagrams

When investigating changes along a transect, often the distribution of a factor is recorded. For example, when studying features such as a sand dune, salt marsh or upland glaciated environment, the percentage cover of different plants can be recorded using quadrats and then plotted on a kite diagram.

To create a kite diagram (see Figure 19), first draw a horizontal line representing the entire length of the transect, using an appropriate scale — the line should not be too short, otherwise the diagram will not give an effective representation. Above this line, draw a parallel line for every different plant species recorded and label each with the species name.

At each survey point along the transect, divide the percentage cover figure for that particular plant species in half. Using a scale appropriate for all the data, plot above the horizontal line to a point representing half the percentage cover figure and the same below the line. Carry on across the length of the transect. For each species, draw a line to connect each point above the central horizontal line and a separate line connecting points below the horizontal line. If the percentage recorded at any point was 0, draw the lines onto the horizontal line.

Exam tip

Make sure you give the full reference including page number for any diagram or graph you have copied from a published source.

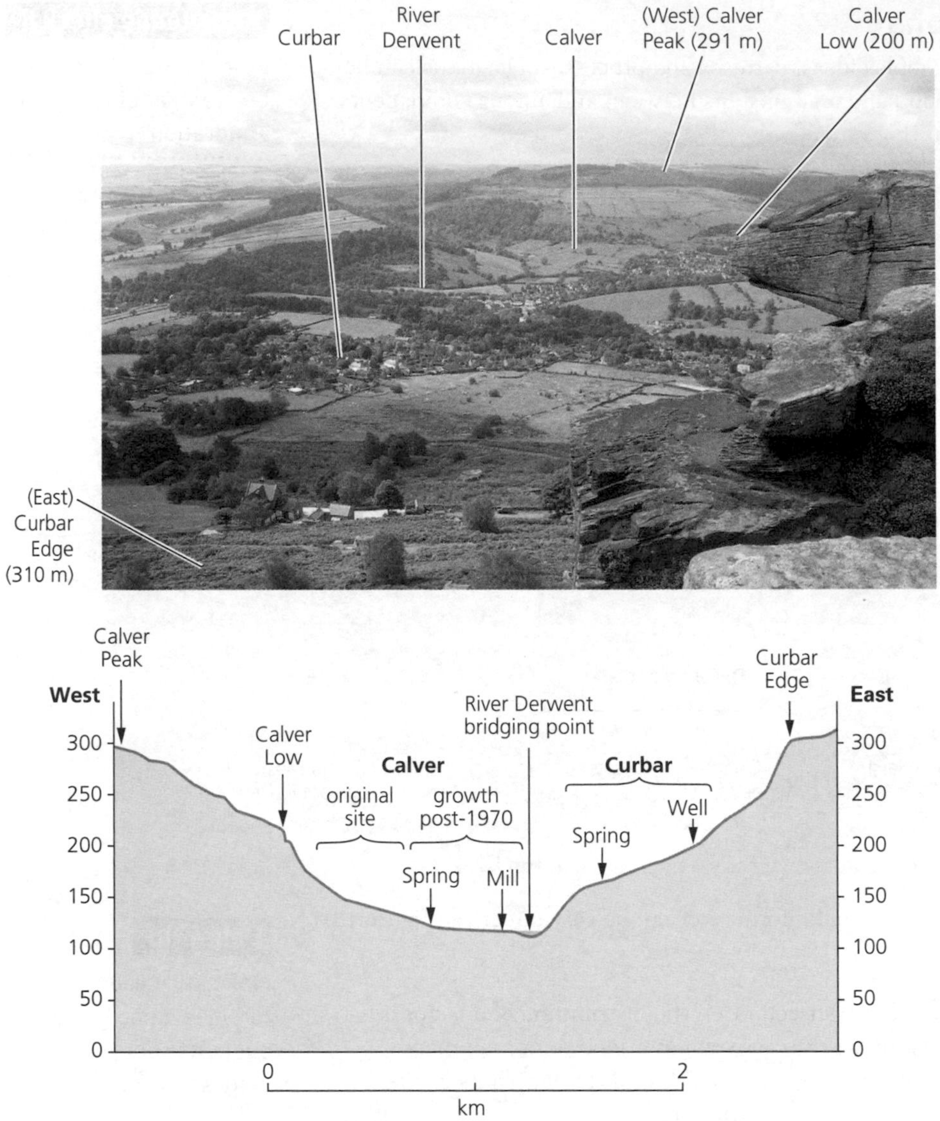

Figure 18 Settlement sites in the Derwent Valley

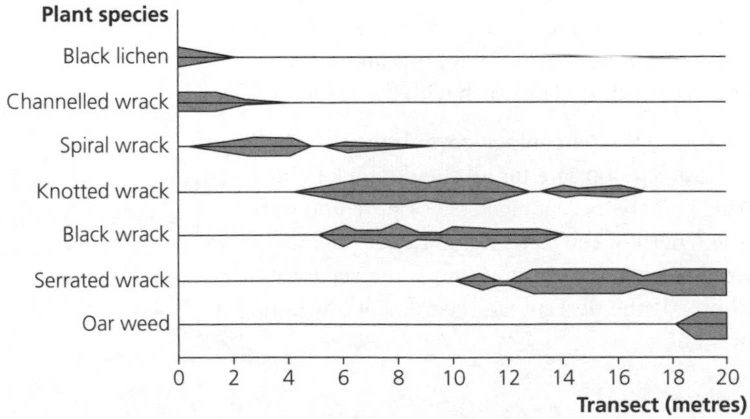

Figure 19 Kite diagram for selected foreshore plants

Graphical skills are widely used in geographical investigation as they offer a means of identifying patterns and trends that may not be clear just by looking at the raw data.

Table 8 Making the most of graphs

Drawing graphs	■ Consider the range in data and choose appropriate scales for the two axes; no data should extend beyond the axes ■ If using a computer package, make sure that you (not the computer) choose the scales ■ Hand-drawn graphs are perfectly acceptable and can be more flexible, e.g. adding labels, and they may be quicker to complete ■ Avoid three-dimensional graphs unless you are representing volume; even then they can be confusing to read
Presenting graphs	■ Give each a figure number, title, and label the axes ■ They should be neither too large nor too small ■ They should integrate with the text

Bar graphs

Sometimes called bar charts, bar graphs present data as a series of rectangles proportional in length (height) to the values of the data. They can be used to show frequencies or percentages. They are useful in presenting changes through time as well as comparing different places. They can be drawn so that the rectangles are horizontal, one example being age–sex pyramids.

Stacked or compound bar graphs use the rectangle to present two or more sets of data and can either use absolute or proportional values such as percentages (Figure 20).

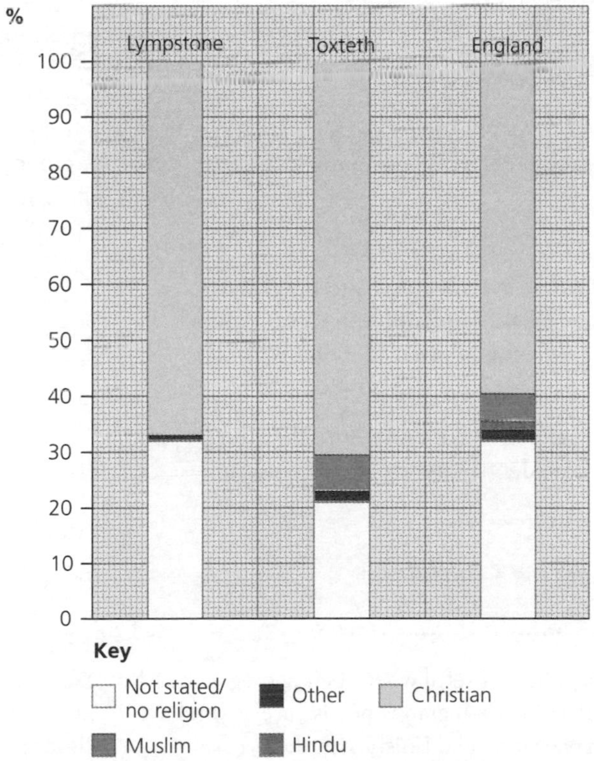

Figure 20 Religious affiliation, Lympstone and Toxteth

Exam tip

Remember to integrate diagrams and graphs with the text. Graphs and charts rarely need to take up an entire page. A single pie chart on a page is not effective presentation of data.

Knowledge check 9

Why can using percentages be useful when comparing two or more sets of data?

Pie charts

These circular graphs can present data in both frequency and percentage forms with the latter being most common. The first and largest segment should start at 12 o'clock with subgroups in the population being drawn often in order of descending value. Pie charts are not suitable for data with fewer than four subgroups but, equally, a large number of subgroups is inappropriate as it becomes difficult to differentiate amongst them. A maximum of seven or eight subgroups is best (Figure 21).

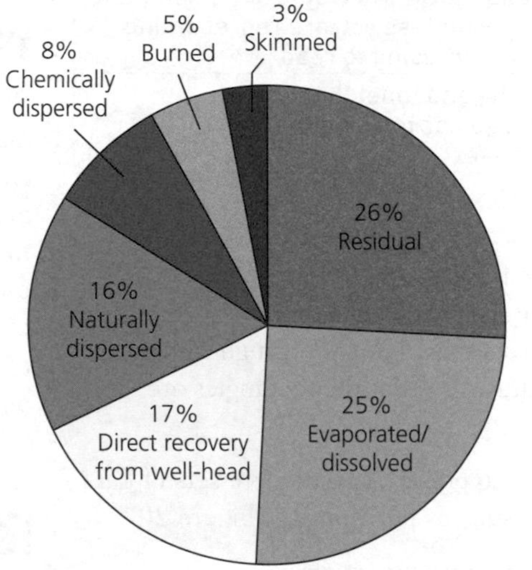

Residual = oil in ocean, washed ashore and collected or in sand and sediments
Naturally dispersed = oil currently being degraded naturally, e.g. by bacteria

Figure 21 Where oil from Deepwater Horizon has gone

Exam tip

Make sure that each segment of a pie chart is clearly labelled with the data it represents.

Line graphs

Line graphs are used to present continuous change in data either through time (e.g. population) or across distance (e.g. average height of plants along a sand dune transect). When plotting, the x axis should be the **independent variable** with the **dependent variable** on the y axis.

Comparing sets of data on a line graph can be effective but make sure that the graph does not become cluttered, making it difficult to identify trends and to read values.

Exam tip

Often in geographical investigation, distance is an important variable. This always goes along the x axis.

Plotting several categories of data one above the other is useful when their aggregate trend and pattern need to be visualised (Figure 22). Reading such graphs needs care as you have to separate the categories and note where each one starts and finishes on the y axis.

The **independent variable** is expected to influence another factor, for example distance from a sports stadium affects litter levels.

The **dependent variable** responds to changes in the other variable, for example pebble size varies in response to distance along a beach.

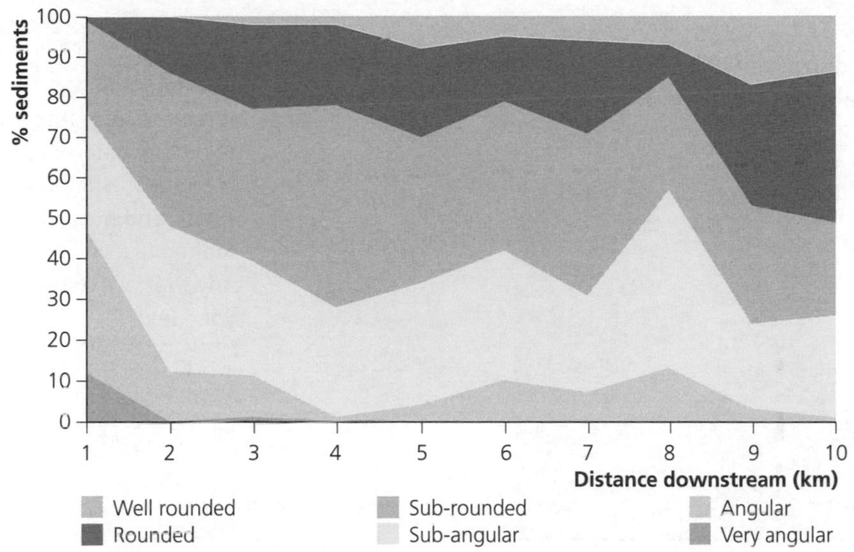

Figure 22 Downstream changes in sediment roundness, River Aire

Triangular graphs

When data can be subdivided into three subgroups and together they add up to 100%, a triangular graph is an effective visualisation of those data. Data such as the relative proportions of age groups in a population (young, adults, elderly) or employment structures (primary, secondary, tertiary) can be convincingly presented on a triangular graph (Figure 23).

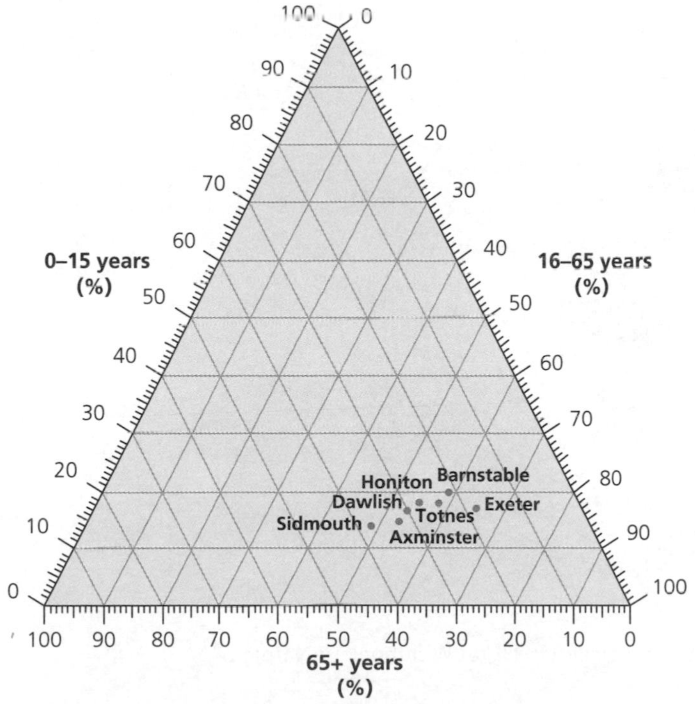

Figure 23 Age structures in Devon districts, 2015

Dispersion charts

These have a single axis, usually drawn vertically. They can be used to compare different data sets and are also valuable in helping decide categories before drawing a choropleth map (Figure 24).

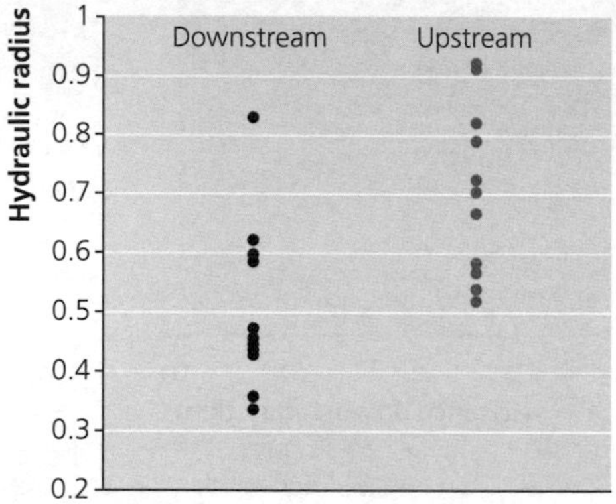

Figure 24 Dispersion diagram showing variations in hydraulic radius, Eglingham Burn, Northumberland

Directional graphs

To represent data about the direction of a factor, such as wind direction (known as a wind rose graph) or orientation of long axes of sediment, a directional graph is useful. It is based on points of the compass and quantifying the frequency of the factor is included (Figure 25).

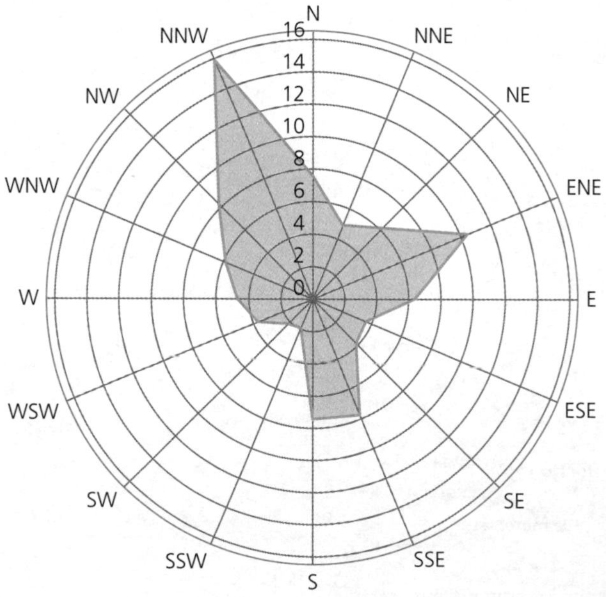

Figure 25 Wind rose showing average wind direction (%) at Windhoek in Namibia

Exam tip

It can be useful to add to a dispersion diagram measures such as mean and median. Make sure you refer to any that are added in your text.

Locating graphs on maps

Placing graphs, especially bar graphs and pie charts, on a map can be a very effective technique to show geographical patterns. It allows you to place data in their actual locations and this can help reveal spatial patterns (Figure 26). This can be achieved by using Digimaps or drawing a base map yourself.

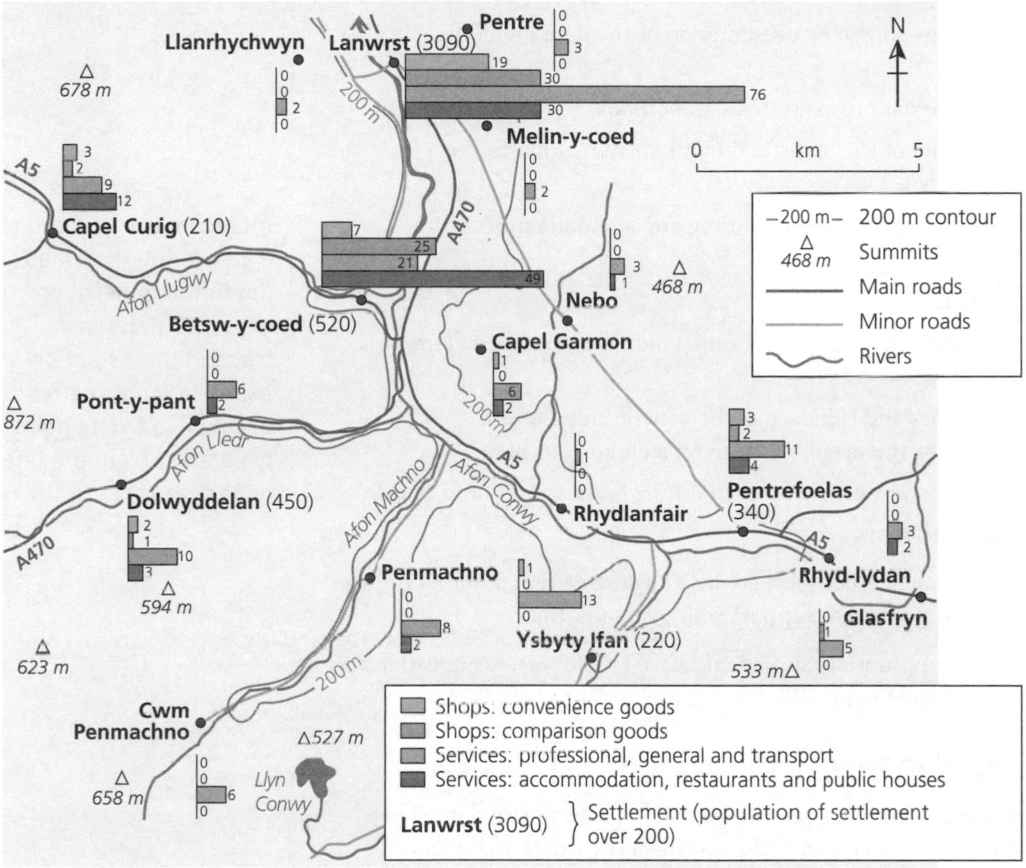

Figure 26 The location of shops and services in the upper Conwy basin, north Wales

Marking criteria

Section 3: Data presentation techniques	
Level 3 (7–9 marks)	■ There is appropriate and selective presentation of the most influential data collected directly related to the investigation. ■ The range of data presentation techniques is appropriate and well-selected, with good knowledge and understanding of the relevant techniques for representing results clearly. ■ There is an appropriate balance of simple and more sophisticated data representation methods relevant to the topic.

Data analysis and explanation

At this stage of the investigation, primary and secondary data collection is complete. It may be that while you are analysing your data, you discover that further secondary data are needed to help with explanation.

Four elements make up this stage:

1. descriptions of the findings shown by presentation of the data including descriptive statistics
2. appropriate analysis of the data using statistical methods
3. interpretation of the results of the statistical methods in relation to the original question/hypothesis
4. explanation of the patterns discovered including any **anomalies**

An **anomaly** (residual) is a data value that does not fit the general or expected pattern or trend in the data. For example, on a scatter graph, it is a data point lying away from the line of best fit.

Describing data

Any data presented in a table, chart, diagram or map must be described. Three aspects should be covered.

1. Describe the main patterns and trends, e.g. the general trend of the points on a scatter graph or the overall pattern on a choropleth map.
2. Give examples of actual data values from the tables, charts, graphs and maps.
3. Pick out the anomalies/residuals.

You should try to achieve a balance between giving a clear, coherent summary of the patterns and trends and giving detail to support your generalisations.

When deciding on the most appropriate statistical test(s) to use, various questions should be asked about the data (Figure 27).

Descriptive statistics

Ratios, fractions, percentages and densities can be used to summarise data. The general nature of a data set is revealed and this can suggest further analytical techniques you could apply.

Exam tip

When using percentage change, remember to always give *n* used in the calculation. Adding 1 to 1 is a 100% increase but adding 1 to 100 is only a 1% increase. The value of *n* is very significant.

Measures of central tendency

A set of data can be summarised by the use of a single value around which all the other data are distributed. These measures can also be helpful in highlighting similarities or differences between data sets.

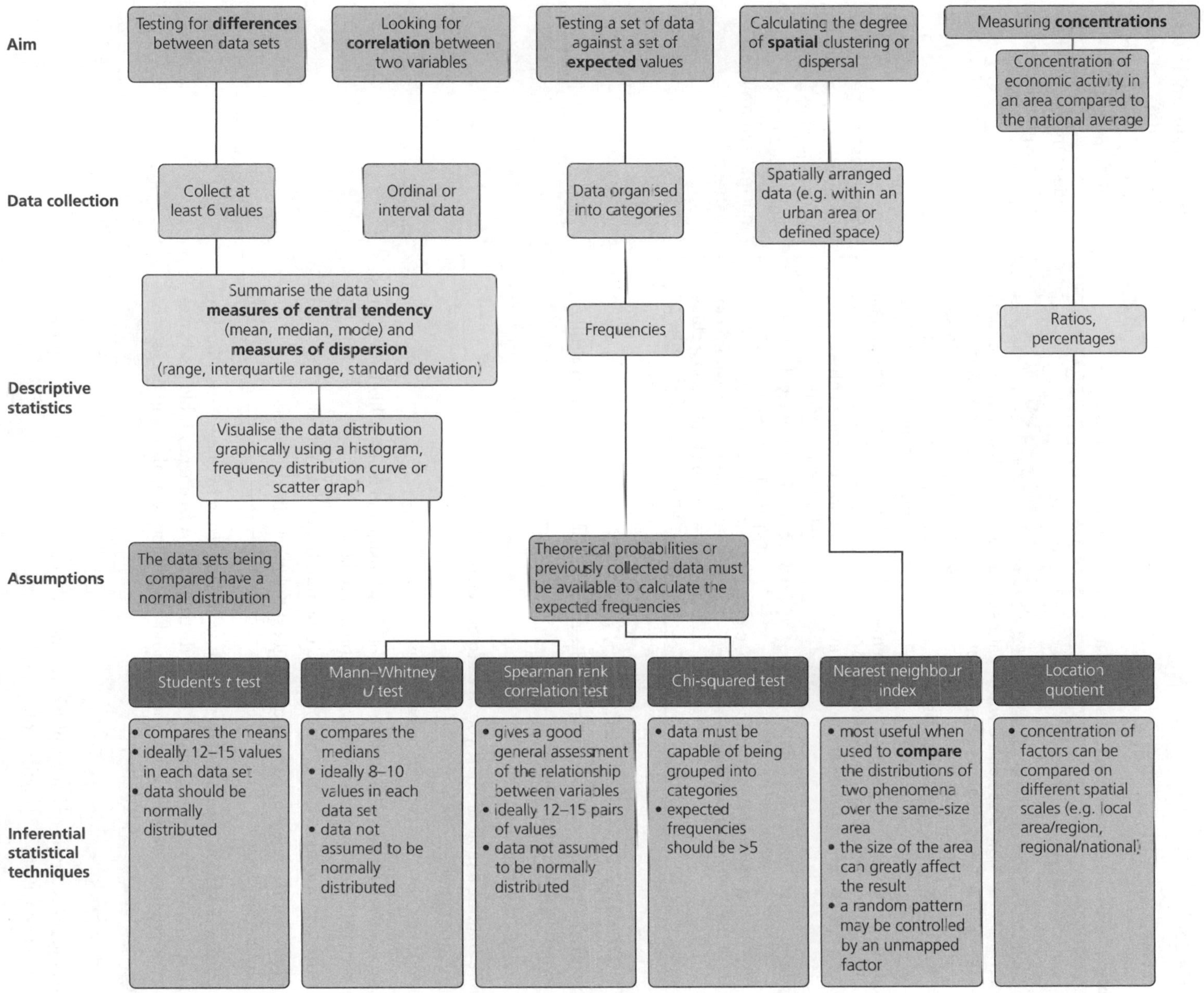
Aim
Data collection
Descriptive statistics
Assumptions
Inferential statistical techniques
Testing for **differences** between data sets
Looking for **correlation** between two variables
Testing a set of data against a set of **expected** values
Calculating the degree of **spatial** clustering or dispersal
Measuring **concentrations**
Concentration of economic activity in an area compared to the national average
Collect at least 6 values
Ordinal or interval data
Data organised into categories
Spatially arranged data (e.g. within an urban area or defined space)
Summarise the data using **measures of central tendency** (mean, median, mode) and **measures of dispersion** (range, interquartile range, standard deviation)
Frequencies
Ratios, percentages
Visualise the data distribution graphically using a histogram, frequency distribution curve or scatter graph
The data sets being compared have a normal distribution
Theoretical probabilities or previously collected data must be available to calculate the expected frequencies
Student's *t* test
Mann–Whitney *U* test
Spearman rank correlation test
Chi-squared test
Nearest neighbour index
Location quotient
• compares the means • ideally 12–15 values in each data set • data should be normally distributed
• compares the medians • ideally 8–10 values in each data set • data not assumed to be normally distributed
• gives a good general assessment of the relationship between variables • ideally 12–15 pairs of values • data not assumed to be normally distributed
• data must be capable of being grouped into categories • expected frequencies should be >5
• most useful when used to **compare** the distributions of two phenomena over the same-size area • the size of the area can greatly affect the result • a random pattern may be controlled by an unmapped factor
• concentration of factors can be compared on different spatial scales (e.g. local area/region, regional/national)

Figure 27 Choosing an appropriate statistical technique

Three measures of data distribution are widely used:

1 **mean** (or arithmetic mean) often called the 'average'

2 **median**

3 **mode**

Mean is the average value in a data set.

Median is the middle value in a data set when values are ranked in order of magnitude.

Mode is the group/class of data with the most values.

Table 9 Advantages and disadvantages of measures of central tendency

Measure of central tendency	Advantages	Disadvantages
Mean	■ Makes use of the whole data set ■ Gives a simple overview of the whole data set ■ Used in many statistical methods	■ Different distributions of data can give similar means ■ May not represent a data set accurately when there are a number of high or low values
Median	■ More representative than mean when data set is **skewed** ■ Straightforward to calculate	■ Different distributions of data can give similar medians ■ Cannot be used in other statistical methods
Mode	■ Useful when data are in categories or classes ■ Straightforward to obtain, e.g. from a bar graph	■ Value of the principal mode depends on the choice of categories/classes ■ Can be ambiguous when there is a range of values in a class (sediment x axis 3–5 cm)

Skewness is the degree to which the distribution of a data set varies from a normal symmetrical pattern.

Frequency distributions

How often a value occurs in a data set is its frequency. These values are usefully presented in a table or a chart.

Frequency tables

Before making a table or chart you need to decide on:

- the number of classes
- the range of values covered by each class

Using too few classes will result in a loss of detail, using too many is likely to confuse the pattern. Two simple calculations can help you decide:

1 **number of classes:** $\sqrt{n}$ when n is the total number of data values

2 **class width:** range in data values (highest to lowest) divided by the number of classes

If your investigation generates large amounts of data, using frequency distributions helps to identify the general shape of the distribution. A frequency table also gives descriptive statistics such as the median and mode. They can also help decide categories for a choropleth map.

Histograms

A histogram is a type of bar graph/chart where the frequency of values is represented in classes or categories (Figure 28).

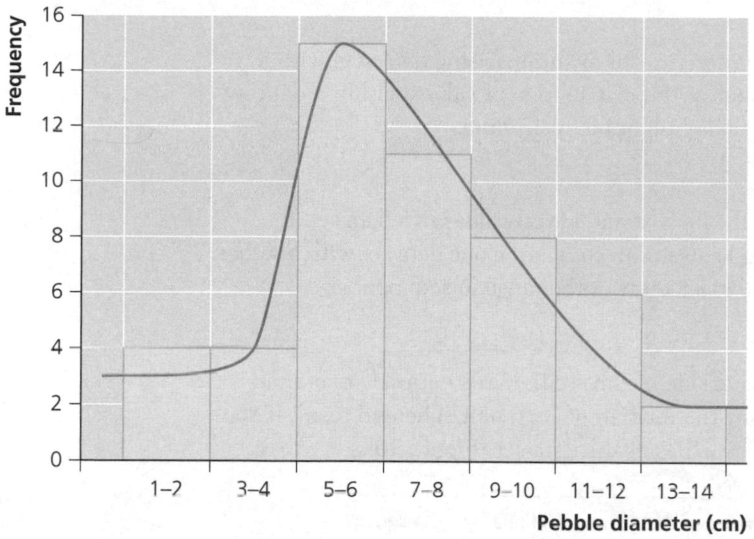

Figure 28 An example of a histogram

Frequency distribution curves

A frequency distribution curve can be added to a histogram to highlight the shape of the distribution (Figure 28). There are three main types of frequency distribution curves: normal, positively skewed and negatively skewed.

Type of curve	Characteristics
Normal 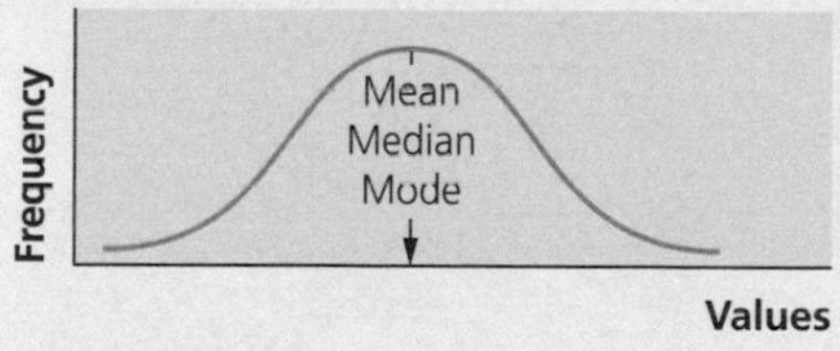	Curve symmetrical; bell-shape; mean, median and mode have same value or will be very close to each other
Positively skewed 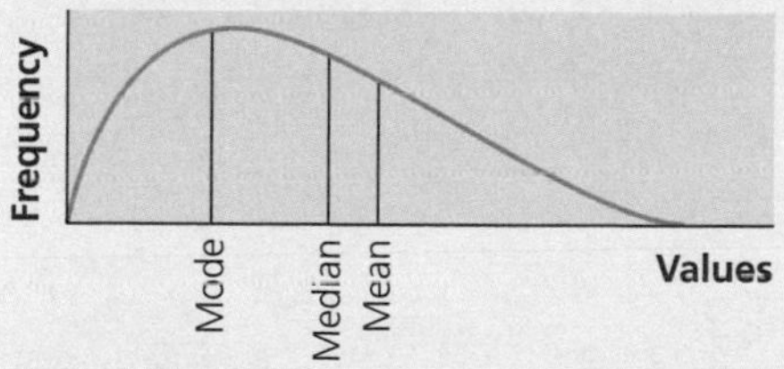	Curve asymmetric; peak on left hand side; longer tail towards right hand side; most values below the mean
Negatively skewed 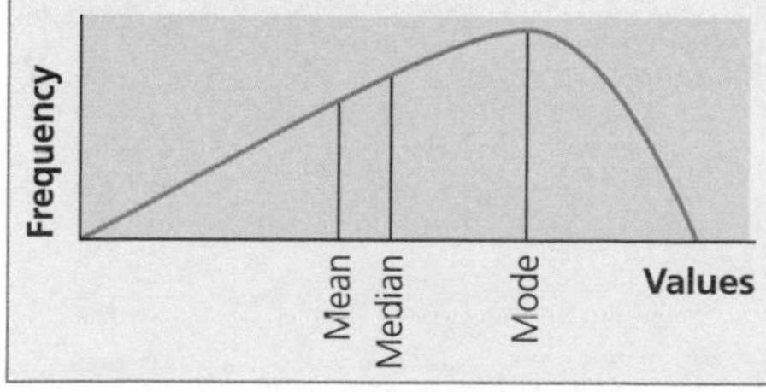	Curve asymmetric; peak on right hand side; longer tail towards left hand side; most values above the mean

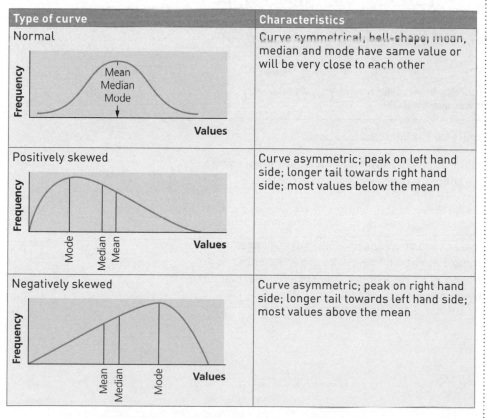

Figure 29 Types of frequency distribution curves

Measures of dispersion

The three measures of central tendency are useful in summarising data sets with a single value. However, it is useful to look at the distribution of values within a data set as the average value (mean, median, mode) might be misleading.

Range

This is simply the difference between the highest and lowest values in a data set. It emphasises the extreme values but can be useful in comparing one data set with another such as the range in distances travelled to a facility or the range in sediment size.

Interquartile range

This is the interval extending 25% either side of the median in a data set. Each half can then be divided into two by finding the median of each half. The end result is the identification of the upper and lower quartiles of the data set (Figure 30).

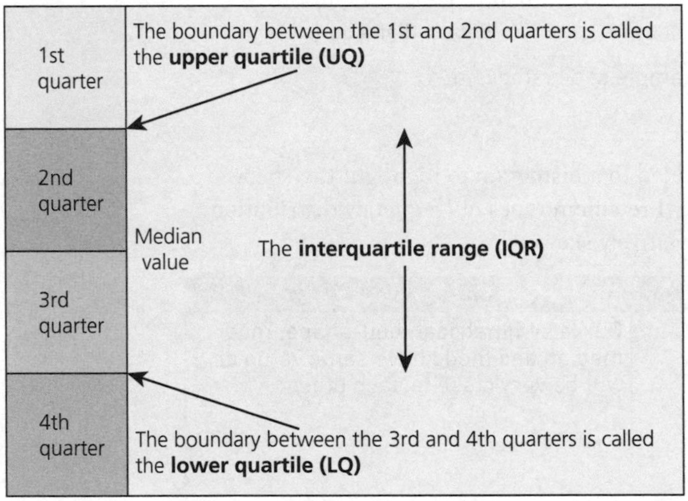

Figure 30 The interquartile range

It can be very helpful to label the upper and lower quartile values when using a dispersion graph as well as any other measures you refer to in the text.

Exam tip

In the AS exam, you may be assessed on mean, median, mode, range, interquartile range and standard deviation. You will be expected either to complete a calculation and/or to interpret a figure. You should understand the purposes and differences between these measures.

Standard deviation

Standard deviation (SD) measures the dispersion of values around the mean. Unlike the range and interquartile range, it uses all the values in a data set. It can be a helpful statistical tool when comparing two data sets. The mathematical formula for standard deviation is:

$$\sigma = \sqrt{\frac{\sum (x-\bar{x})^2}{n}}$$

where:

σ (sigma, the Greek lowercase letter 's') is 'standard deviation'

Σ (Sigma, the Greek capital letter 'S') is 'sum'

x is each value in the data set

$\bar{x}$ is the mean value of the data set

n is the number of items in the data set

In other words, add up the squared deviations of all the data values from the mean, divide by the total number of data values, and then take the square root.

The size of the standard deviation indicates the degree of dispersion from the mean value. When the distribution of values in a data set is close to normal, just over two-thirds (68.27%) of the values lie between plus or minus one standard deviation of the mean and 95.45% lie between two standard deviations.

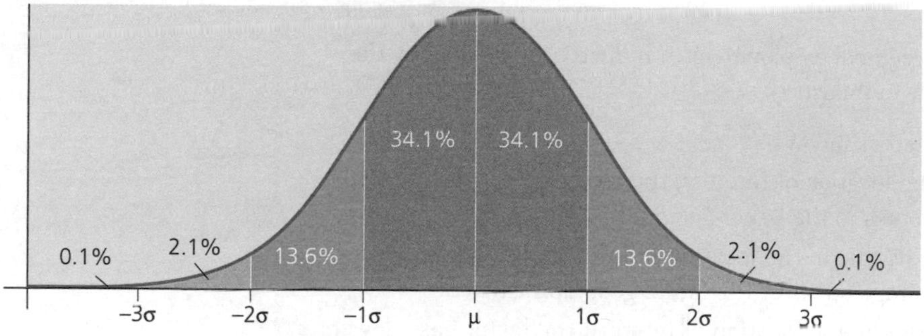

Figure 31 Normal frequency distribution

Coefficient of variation

The coefficient of variation (CV) is used to compare data sets with very different means. It measures the standard deviation as a percentage of the mean and is another way of indicating the degree of dispersion of the data set.

$CV = (\sigma/\bar{x}) \times 100$

This statistic can be useful when plotting the variability of data on a map as percentages can be readily compared.

Measures of concentration

Geographers focus on spatial patterns in many investigations and are often interested in how concentrated or dispersed a factor is.

Exam tip

In the AS exam you will be given the formula for standard deviation if necessary, so you do not need to learn it. However, you do need to know how to apply and interpret it.

Exam tip

It is helpful to remember the percentage of values that lie within plus or minus one SD (68%), two SDs (95%) and three SDs (99%) for the AS exam as you will be expected to interpret outcomes of statistical tests.

Location quotient

This measures how concentrated a factor is within a certain area, such as the distribution of particular demographic or socio-economic characteristics, for example occupations, age or ethnic groups in wards across a city or districts across a county. It can also be useful for comparing a region with the national scale.

A location quotient (LQ) is an index number, defined using fractions or ratios. For instance, if employment is the factor being studied, the LQ is given by the formula:

$$LQ_i = \frac{e_i/e}{E_i/E}$$

where:

LQ_i denotes the location quotient for industry i in the regional economy

e_i is employment in industry i in the regional economy

e is total employment in the region

E_i is employment in industry i in the national economy

E is total employment nationally

When the ratios e_i/e and E_i/E are expressed as percentages, the formula can be simplified to:

$$LQ_i = \frac{\% e_i}{\% E_i}$$

In other words, the percentage regional employment in industry i is divided by the percentage national employment in industry i.

Location quotients are interpreted as follows.

- A value of **1** means that in the location of interest, the factor is represented with exactly the same concentration as in the larger geographical area.
- A value **greater than 1** means that in the location of interest, the factor has a higher concentration than average (i.e. over the larger geographical area).
- A value **less than 1** means that in the location of interest, the factor has a lower than average concentration.

Nearest neighbour index (NNI)

This method was developed by botanists studying the distribution of plants. It has applications for investigations into spatial distributions of factors such as comparing patterns of city centre functions (e.g. financial services, types of retailing) or village locations.

The NNI is usually denoted by 'Rn', and the formula used to calculate it is:

$$Rn = 2\bar{d}\sqrt{\frac{n}{A}}$$

where:

$\bar{d}$ is the mean distance between each point and its nearest neighbour.

n is the total number of points in the survey

A is the study area

Rn values can range from 0 to 2.15, and have the following interpretations.

Rn = 0 — clustered: a pattern where the points are grouped closely together. For example, specialist shopping areas such as the jewellery quarter in Birmingham will exhibit this kind of pattern.

Rn = 1.0 — random: points are randomly distributed throughout the area.

Rn = 2.15 — regular: a pattern that is perfectly uniform with regular spacing, such that all the points are at an equal distance from each other. In reality this rarely occurs; the nearest examples might be the settlement pattern in areas such as East Anglia, or parts of mid-western USA.

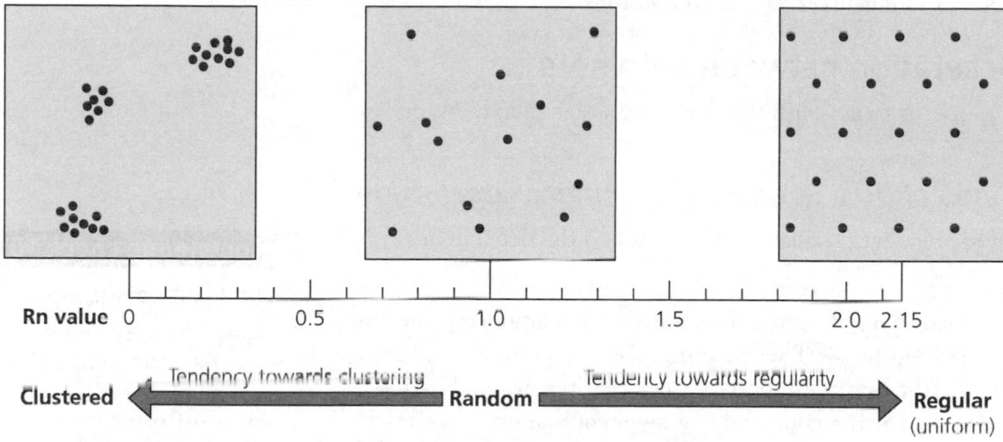

Figure 32 Nearest neighbour index values and their interpretations

There are some things to consider if you are going to make effective use of the nearest neighbour index.

- It is best used when comparing distributions, not just one pattern.
- Each data set should cover a very similar area.
- Be aware of the issues around setting of boundaries to the area being studied, such as those that depend on the choice of the researcher.
- You should consult a critical values table when assessing the significance of your result. This helps you decide if significant clustering or dispersal is present.
- You must appreciate that the index cannot distinguish between a single cluster or a multi-clustered pattern (Figure 33).

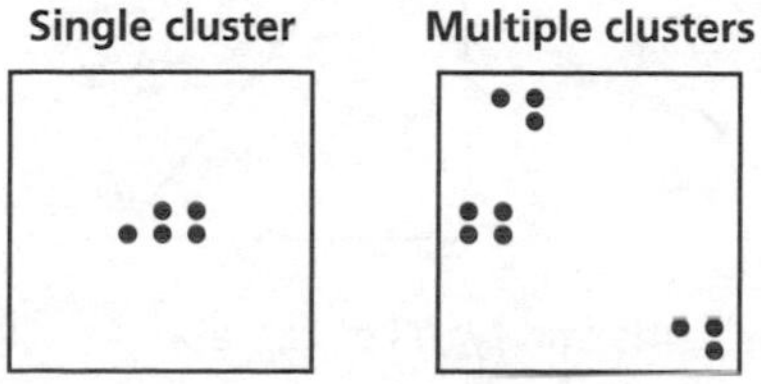

Figure 33 A single cluster and a multi-clustered distribution

Analysis of data

Once you have collected, presented and described your data, it might be possible to apply various quantitative or statistical techniques to analyse them.

Inferential statistics

Because your investigation will almost certainly rely on samples rather than the whole population, there is a possibility that the results are due to chance. Inferential statistical tests help by calculating the probability that the results are due purely to chance. If that probability is low enough, you can be confident that the results are suggesting something worth analysing and explaining.

By setting up a null hypothesis first, your statistical analysis has a sharp focus and you are less likely to forget what the purpose of the testing is. The null hypothesis states: *There is no significant relationship between variable X and variable Y.*

Measures of correlation between data sets

The focus of these techniques is to establish the degree of association between two data sets or variables.

Scatter graphs, lines of best fit and simple linear regression

Scatter graphs plot the independent variable on the x axis and the dependent variable on the y axis.

A line of best fit can be added to highlight the direction of the relationship suggested by the scatter of data points (Figure 34). There should be an equal number of points on either side of the line and the line does not have to start at the origin. Scatter graphs are useful in suggesting the direction and degree of correlation (association) between two variables, as well as highlighting any anomalies in the data set.

Knowledge check 10

What is meant by the 'dependent' variable and the 'independent' variable?

Exam tip

If you are using a computer program to generate scatter graphs make sure that the choice of scale along the axes is suitable. Very often, a hand-drawn graph is more effective and might allow labelling and annotation to be more effective.

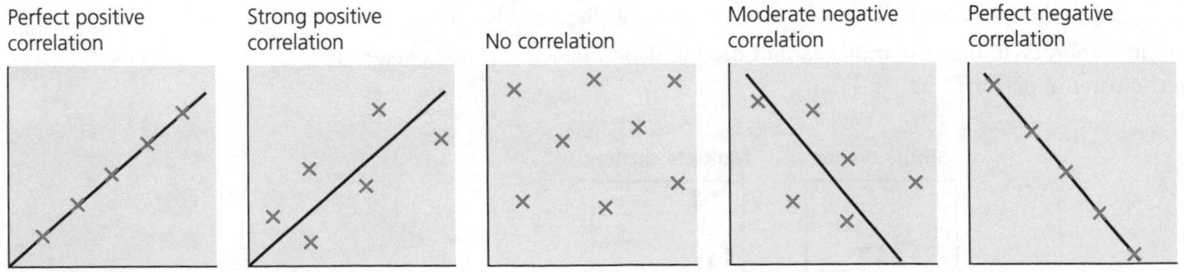

Figure 34 The relationships between scatter patterns and types of correlation

Simple linear regression allows you to plot accurately a line of best fit (Figure 35). It can be achieved relatively straightforwardly by using an Excel spreadsheet (see pp. 543–44 in the *OCR Geography* textbook published by Hodder Education). Regression analysis has two important uses:

1 it allows the prediction of the y value when the x value is known
2 the regression equation gives a precise model of the relationship between the two variables and so allows comparisons with the same variables in different locations

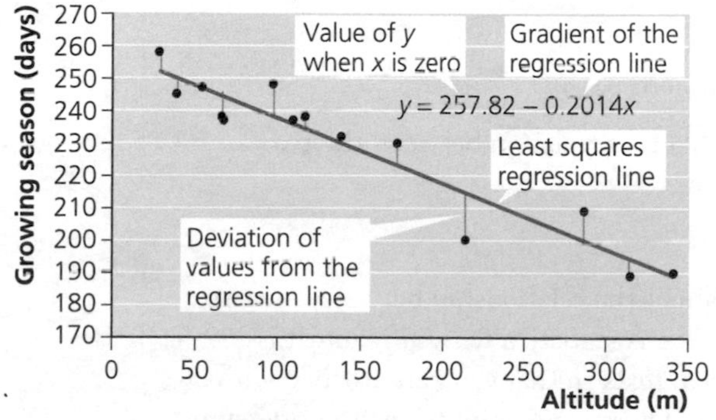

Figure 35 Linear regression for length of growing season and altitude in northern England

Spearman's rank correlation coefficient

If plotting data on a scatter graph suggests that the two variables have a relationship worthy of further investigation, i.e. the scatter of points is not too dispersed in either a positive or negative direction, then you should apply a correlation technique. This will allow you to test how strong the relationship is between the two variables and if it is statistically significant, or whether the relationship could occur simply by chance. For example, analysis of the relationship between per cent vegetation cover and moisture content of soil across a sand dune system.

You can use Spearman's rank with any raw data or percentages as the data do not have to have a normal distribution. It is best if you have at least ten pairs of data but no more than 30.

The formula for Spearman's rank correlation coefficient is:

$$R_S = 1 - \frac{6\sum d^2}{n^3 - n}$$

where:

d is the difference in rank order of each pair of data values

n is the number of pairs of data values

Σ (Sigma, the Greek capital 'S') is the sum of all the d^2 values

The two data sets are ranked. It does not matter whether you rank from highest value to lowest or lowest to highest. It is vital, however, that you rank both sets of data in the same way. If your data sets have several tied ranks, the value of this statistical

Exam tip

The analysis of data using statistics can be greatly helped by the use of Excel. This package includes most of the statistical tests that you are likely to use.

test is reduced. A tied rank is when each data set has values that have the same rank. This means that *d* is 0

The R_S value is usually calculated to two decimal places and is interpreted as follows.

A positive value indicates positive correlation: as one factor increases, so does the other. A value of +1 indicates perfect positive correlation — the data points lie exactly on a line with positive slope.

A negative value indicates negative correlation: as one factor increases, the other decreases. A value of –1 indicates perfect negative correlation — the data points lie exactly on a line with negative slope.

A value of 0 (or close to 0) indicates no correlation.

The closer R_S is to +1 or –1, the stronger the likely correlation between the two factors.

Testing for significance

Although your R_S result tells you something about the relationship between the two data sets you cannot be sure that the result alone is statistically significant. It is necessary to test for significance with statistical tests so that you can proceed with your analysis and explanation with confidence. There is no point trying to understand and draw conclusions if your results are not significant. It might be that chance is the key factor!

Significance testing uses the size of the data sets to obtain the critical values from tables of values which are widely available.

For your investigation the significance level most likely to be appropriate is 0.05. This means that there is only a 5 in 100 likelihood of the results having occurred by chance or, in other words, that for 95 out of 100 repeats of the investigation the result would be the same. Achieving this level of significance allows you to tackle explaining the results with a great deal of confidence.

Pearson's product moment correlation

Spearman's rank correlation is widely used in geography but there are other tests available. Pearson's test is similar to Spearman's in that it generates a correlation coefficient but it should only be used when your data have a normal distribution. It is available in Excel (see p. 543 in the *OCR Geography* textbook).

Measures of differences between data sets

Geographical investigation often looks at differences between samples. As for correlation, several statistical tests are available to analyse differences between samples.

Student's t-test

You can compare two data sets by looking at their means but, for an objective analysis, the Student's *t*-test is useful. Perhaps its primary value is that it indicates the extent to which any difference could have occurred just by chance. It should only be used when your data set has a normal distribution and when the standard deviations are equal.

Exam tip

While the formula may look complicated, following the worked example in the *OCR Geography* textbook on page 537 will help you apply the test to your own data.

$$t = \frac{|\bar{x} - \bar{y}|}{\sqrt{\dfrac{\dfrac{\sum x^2}{n_x} - \bar{x}^2}{n_x - 1} + \dfrac{\dfrac{\sum y^2}{n_y} - \bar{y}^2}{n_y - 1}}}$$

where:

$\bar{x}$ and $\bar{y}$ are the means of the two data sets

n_x and n_y are the numbers of values in each data set

Σ (Sigma, the Greek capital 'S') is the sum of all x/y^2

Note that $|\bar{x} - \bar{y}|$ means the 'absolute difference' between $\bar{x}$ and $\bar{y}$; it is the **size** of the difference between $\bar{x}$ and $\bar{y}$, and is always positive.

Mann–Whitney U test

This test also assesses the significance of difference between two data sets. However, unlike the Student's t-test, it can be used when the data sets do not have normal distributions. This makes it a very valuable analytical tool for geographers as quite often statistical populations are skewed, for example pebble size at either end of a beach or crop yields per hectare from farms in different locations.

Other advantages include:

- it can be applied to small, medium and large data sets
- it can be applied to data sets containing unequal numbers of values
- it can be applied to ordinal data — data placed in an order, e.g. highest to lowest

To calculate the U values for both data sets the formulae are:

$$U_x = n_x \times n_y + \frac{n_x(n_x + 1)}{2} - \sum r_x$$

$$U_y = n_x \times n_y + \frac{n_y(n_y + 1)}{2} - \sum r_y$$

where:

n_x and n_y are the numbers of values in each data set

r_x and r_y are the ranks of the values in each data set

If you decide to use the Mann–Whitney U test, it is probably best that you first read through the worked example on pp. 538–39 in the *OCR Geography* textbook. This should give you the confidence to apply the test to your own data.

Chi-squared test

This test compares observed and expected (theoretical) distributions of data to assess whether or not there is a significant difference between them. For example, assessing whether there is any significant difference in perceptions of a rebranding scheme by different age groups or genders. There are two versions of the test: a one-sample test and one when there are two or more samples to be compared.

Exam tip

There are worked examples of both versions of the Chi-squared test in the *OCR Geography* textbook on pages 539–40.

The Chi-squared test is another powerful test as it also does not require the data to be distributed normally. However, there are some conditions that need to be met before you can apply it:

- data must be organised into categories
- data must be in frequencies not in percentages nor proportions
- there should not be too many categories for which expected frequencies are small
- the total number of observed data values must be greater than 20
- the expected frequency of each category should be greater than 4

The formula for the chi-squared value is:

$$\chi^2 = \sum \frac{(O - E)^2}{E}$$

where:

O is observed data value

E is expected frequency

Σ (Sigma, the Greek capital letter 'S') is 'sum'

The Chi-squared test is used to compare an observed distribution of something with an expected distribution. The expected distribution assumes that whatever is being studied, such as the distribution of pebbles of a particular shape on a beach or the distribution of post offices with increasing distance from a town centre, is the same across the beach or town.

The Chi-squared test will indicate if there is a significant difference in the distribution.

Explanation of data

The findings of your investigation will have revealed patterns in the data. Some of the patterns will have been obvious in the various presentation techniques you used. Others will have required a statistical test to confirm that a statistically significant relationship or difference exists. For investigations using qualitative and non-numerical analysis, patterns and relationships or differences will also have emerged.

Knowledge check 11

Why is it important to apply a test of significance to statistical analysis?

That is not the end of the matter as you now need to return to the real world and seek reasons for the result. You should re-read your introduction and, in particular, any wider theories, ideas or concepts you mentioned. This should help you focus on the likely cause-and-effect relationships between the factors you have investigated.

Even if the overall pattern or trend in your data is clear, you may have found some anomalies. You should identify these and suggest reasons why they may not fit the general findings. Such reasons may include one or other of the following.

- Spatial patterns often involve many variables and it may be that further research is required to look at one or more variables you had not considered.
- A particular aspect of the real world at the location and time of your data collection has influenced the data, such as weather conditions.
- A problem with some aspect of data collection such as sampling accuracy or equipment malfunction.

If your analysis does not give a clear result, such as not finding a significant difference between two data sets or locations, it is important to appreciate that your investigation has still been worthwhile and should not be dismissed. Your evaluation might suggest a different question or sampling framework is appropriate, for example.

A key aspect of explanation is thoughtful and considered argument backed up with references to your data.

Marking criteria

Section 4: Data analysis and explanation	
Level 4 (11–14 marks)	■ Data and information collected is analysed and interpreted in an effective and coherent manner with evidence of independence, demonstrating the knowledge and understanding of the techniques appropriate for analysing and explaining data and information. ■ When appropriate to the topic, statistical analysis and significance testing are used accurately and proficiently for both data and topic of investigation. ■ When appropriate to the topic, qualitative and non-numerical analysis techniques are successfully and individually developed and used to support explanations and findings from data and information collected. ■ The analysis and explanation are relevant and link effectively to the stated aims or questions or hypotheses. ■ There is effective use of appropriate knowledge, theory and geographical concepts to help explain findings.

Conclusions and investigation evaluation

Because the conclusion section is the last to be written, the temptation is to rush through it. Hopefully you have managed to keep to your timetable, in which case there should be time to consider the conclusion carefully and thoughtfully.

Conclusions linked to the original questions or hypotheses

The final stage in your investigation must not be a separate stage. It must be intimately linked with the purpose of the investigation, which is to answer a question or to test an hypothesis. To achieve this the following steps should be taken:

- read through the introduction before writing the conclusion
- remind yourself of the theories and models relevant to your investigation
- look over the data presentation section to remind yourself of the patterns that emerged
- read through your analysis and explanation section to remind yourself of the key findings

As you go through these stages, note down the key points that strike you. These can then form the basis of your conclusions.

The conclusion may seem to almost write itself, but it doesn't. If you have offered substantial and authoritative work throughout the previous sections then anyone reading the investigation will have a fairly good idea of the direction and destination of your work. However, that is no excuse for a poor conclusion because this offers something different. It is important to generate a convincing conclusion.

- Do not simply repeat the original question/hypothesis: 'Therefore it can be concluded that the rebranding of X has been successful or that the distribution of Y is influenced by several factors'.
- Do not introduce points or material not previously included. If you do think of something new, rewrite the appropriate earlier section to include it there.
- Do not conclude something that you have not focused on. Keep asking yourself 'Have I provided the evidence to support the points in my conclusion?'.
- Do mention further questions that your research suggests might be interesting and of value to investigate. These might refer to some aspect of the theory or model which, quite rightly, your particular investigation did not focus on. It might be that in the location(s) you investigated, some particular local feature was apparent that would be worth further investigation.

Evaluating data, their collection and possible improvements

No research is perfect. For example, limitations exist in the resources available, such as time and equipment. Recognising areas of the investigation that did not go as well as you had hoped and suggesting possible improvements is a sign of mature appraisal.

Reliability and accuracy

A key area to evaluate is the reliability and accuracy of data sources. What is the provenance of secondary material? Make sure you have recorded who published it, where and when. Even with reputable sources such as British universities and the Census, information can be less reliable than you would hope if information relates to some time ago. For example, the Census is conducted every ten years, the most recent being 2011. Updates are available but they are estimates and, although these are likely to be close to the actual levels, nevertheless the issue of reliability should be raised.

Exam tip

It is important that you understand the difference between reliability and accuracy as well as their respective roles in geographical research.

The reliability and accuracy of your primary data are closely related to your sampling strategy. This will directly impact the quality and quantity of data. How effectively did you use equipment in the field? How precise were you in measuring slope angle? Did you stick with your sampling arrangement in choosing who to ask to complete questionnaires or, in the light of many rejections, did you change to asking anyone who you could persuade to stop and talk to you?

Suggesting improvements

Comments about possible improvements are not admissions of failure. They represent mature reflection on where, how and why changes to an area of your investigation might strengthen your research and result in higher standards of reliability and accuracy.

Notes made during primary data collection will remind you of issues that you faced and allow sensible suggestions for improvements. Taking and then annotating photographs of issues affecting data collection can be a very valuable means of communicating your evaluation.

It is crucial that you are sensible in your suggestions. Simply doubling the number of sample points or interviews does not necessarily lead to a significant gain for the investigation. It may be that you collected too few data but this should have been identified at the planning stage. Perhaps the way you set up your sampling framework meant that vital data were missed. It may be that you didn't pay sufficient attention to some aspect of the theory which meant that information was not collected about a vital element.

Suggesting extensions to the investigation can be helpful. Some features change seasonally (beaches, vegetation, visitor numbers, traffic flows, the 'character' of a place) so recognising that you took a 'snapshot' of the pattern and that this may not be the whole picture is a strength. The patterns you observe are not just the result of processes occurring on the day you collected data. Consider what processes might be operating at different times of day, days of the week or months of the year. A survey of the sense of place of an area will give contrasting results if your data collection coincides with collection day for household rubbish, and bags and bins are lining the street ready to be picked up. Building work may give an area a sense of congestion, noise and dust. However, this is temporary and may result in significant improvements. Skips outside houses indicate refurbishment of existing buildings and will not be there forever.

Much of the built environment was constructed under different socio-economic and political circumstances to today and you should be aware of these.

Validity of analysis and conclusions

Evaluate the outcomes of your analysis. How useful were the particular techniques you employed in describing and analysing the data? Statistics reveal but they can also conceal. What help were they in answering your original question or testing your hypotheses?

Can you justify your conclusions on the basis of the analysis? Remember that you are investigating a specific place and issue. Your work has been influenced by particular circumstances. In these ways you need to show that your geographical knowledge and understanding have been extended.

Exam tip

Reference to past patterns and processes is often important in understanding the present-day geography. There is a vast amount of material relating to what was happening previously and you may need tenacious detective work to unearth it.

Marking criteria

Section 5: Conclusions and investigative evaluation	
Level 4 (10–12 marks)	■ There are clear, accurate and thorough conclusions linked to the aims or questions or hypotheses, communicated by means of extended writing. ■ Draw effectively on primary and secondary evidence and, where appropriate, theory to provide a very well argued case and shape conclusions. ■ There is convincing evidence that conducting the investigation extended geographical understanding with clear reference to the wider geographical context of the investigation. ■ There is a strong evaluation of the overall success of the investigation with reference to the reliability of data sources, data collection methods (including sampling), the accuracy of data collected and the extent to which it is representative, and the validity of the analysis and conclusions. ■ There is thorough understanding of the ethical and socio-political dimensions of field research and data presentation.

Overall quality and communication of written work

Language

Write your report in continuous prose that communicates clearly and precisely all aspects of your investigation. Research projects use technical terms that convey knowledge and understanding. Define these terms where necessary, making use of textbook glossaries and specialist dictionaries.

Check your punctuation and spelling. Use — but do not rely on — a spell checker. *To many errors can be mist by students not reading threw there work.*

When you read through your work, check that you do not over-use certain words. Find synonyms or re-write the sentence if there is repetition. Shorter sentences communicate more effectively than long ones.

Use well-known and established acronyms and abbreviations, such as AONB. Give the full definition with the shortened form in parentheses after it the first time you use an acronym or abbreviation. *The Blackdown Hills was made an Area of Outstanding Beauty (AONB) in 1991.* Abbreviations such as 'ha' for hectare are widely used and accepted.

Structure

The OCR marking criteria are organised into a five-section structure and this can be a helpful framework for your report. The sections follow a logical progression through an investigation. Make good use of sub-headings within each section, for example for different data collection or statistical techniques.

Use paragraphs regularly and frequently. Each paragraph should contain just one idea and its supporting material. The challenge is to make the direction of your argument in your report clear to the reader.

Integrating text and figures

Rather than placing figures and tables on separate pages, integrate them in the text. This will help the flow of your report by allowing the reader to link easily text and figures. Word processing the report makes this task relatively straightforward. Figures should be given consecutive numbers and a heading. A really well-presented report will include a list of these, with their page numbers, as part of the Contents section at the start of the report.

Exam tip

Remember to reference actual page numbers and/or individual chapters of sources such as magazine articles and textbooks.

References

Whenever you have quoted directly from a source, be that a traditional printed source or online, a clear and detailed reference must be given in the text. Sometimes a reference is required at the end of a paragraph if you have summarised an idea or finding.

From the very beginning of your investigation, keep a record of all material you have consulted from right across the range of media that you have consulted so that when it is time to compile the **bibliography**, the task is straightforward. This includes any diagrams or photographs you have used. Make sure you include date of publication and, for online sources, the date when you accessed the source.

A **bibliography** is a list of all the resources you have consulted. It is usually organised alphabetically by author.

Appendix

It can be useful to include an appendix section for raw data or supplementary material. However, do not cram in large quantities of material and make sure it does not contain information that should be in the main body of the report.

Marking criteria

Section 6: Overall quality and communication of written work	
Level 3 (7–10 marks)	■ There is a high standard of communication that is relevant to the geographic purpose of the investigation. ■ Arguments are clear, demonstrating a strong degree of individuality. ■ Written work is very well structured, logical, concise, and includes good presentation with text and figures appropriately integrated. ■ Sources and literature references are clearly stated and accurately referenced throughout the investigation. ■ Geographical terminology is technical, used appropriately, and written language errors are rare.

Questions & Answers

Fieldwork: AS Landscape and place

Assessment overview

Fieldwork is an essential element of AS Landscape and place. You need to acquire and develop fieldwork skills related to processes in both physical and human topics. Fieldwork should be an integral and fundamental part in the study of your chosen *Landscape system* and in the study of the Changing spaces; Making places topics.

In the Landscape and place examination, fieldwork skills are assessed in one compulsory question in Section C, worth 24 marks in total. This is made up of two parts:

- part (a) is composed of a few short-answer questions based on an unseen resource, worth a total of 12 marks
- part (b) comprises one longer question based on the skills you have developed through your fieldwork investigation and is worth 12 marks

In the Landscape and place examination paper as a whole, the requirement is:

- one Landscape systems question – a choice of Coastal, Glaciated or Dryland landscapes (29 marks)
- the compulsory Changing spaces; Making places question (29 marks)
- the compulsory Fieldwork question (24 marks)

The overall length of the paper is 1 hour 45 minutes, therefore it is recommended that you spend approximately 30 minutes on the Fieldwork question.

The necessary fieldwork skills should involve you in:

- identifying appropriate field research questions
- understanding how to observe and record data in the field
- devising practical approaches and being able to justify these
- undertaking practical field methodologies to investigate core human and physical processes
- implementing appropriate methodologies to collect good quality data
- applying your knowledge of concepts to understand field observations
- writing a coherent analysis of your findings and justifying your conclusions
- evaluating and reflecting on fieldwork investigations

In addition, you should be able to apply geographical skills related to fieldwork investigations such as understanding:

- what is meant by geographical data and the implications of its collection, study and presentation for communities
- the use of different types of geographical information such as quantitative, qualitative, primary, secondary, numerical and spatial data

- critical questioning of sources of data and information, ways in which data are analysed and presented and their accuracy and reliability
- collection, presentation and analysis of geo-located data including use of Geographical Information Systems
- collection, interpretation and evaluation of qualitative sources and techniques
- quantitative skills, including the application and purpose of selected descriptive and analytical statistical techniques, measurement and sampling

About this section

The questions below are typical of the style and structure that you can expect to see in the Fieldwork question in the AS paper. Each question is followed by examiner comments, preceded by ⓔ, which offer some guidance on question interpretation. Student responses are provided, with detailed examiner comments on each answer, preceded by ⓔ, to indicate the strengths and weaknesses of the answer and the number of marks that would be awarded. A final summary comment is also provided giving the total mark and level.

Question 1

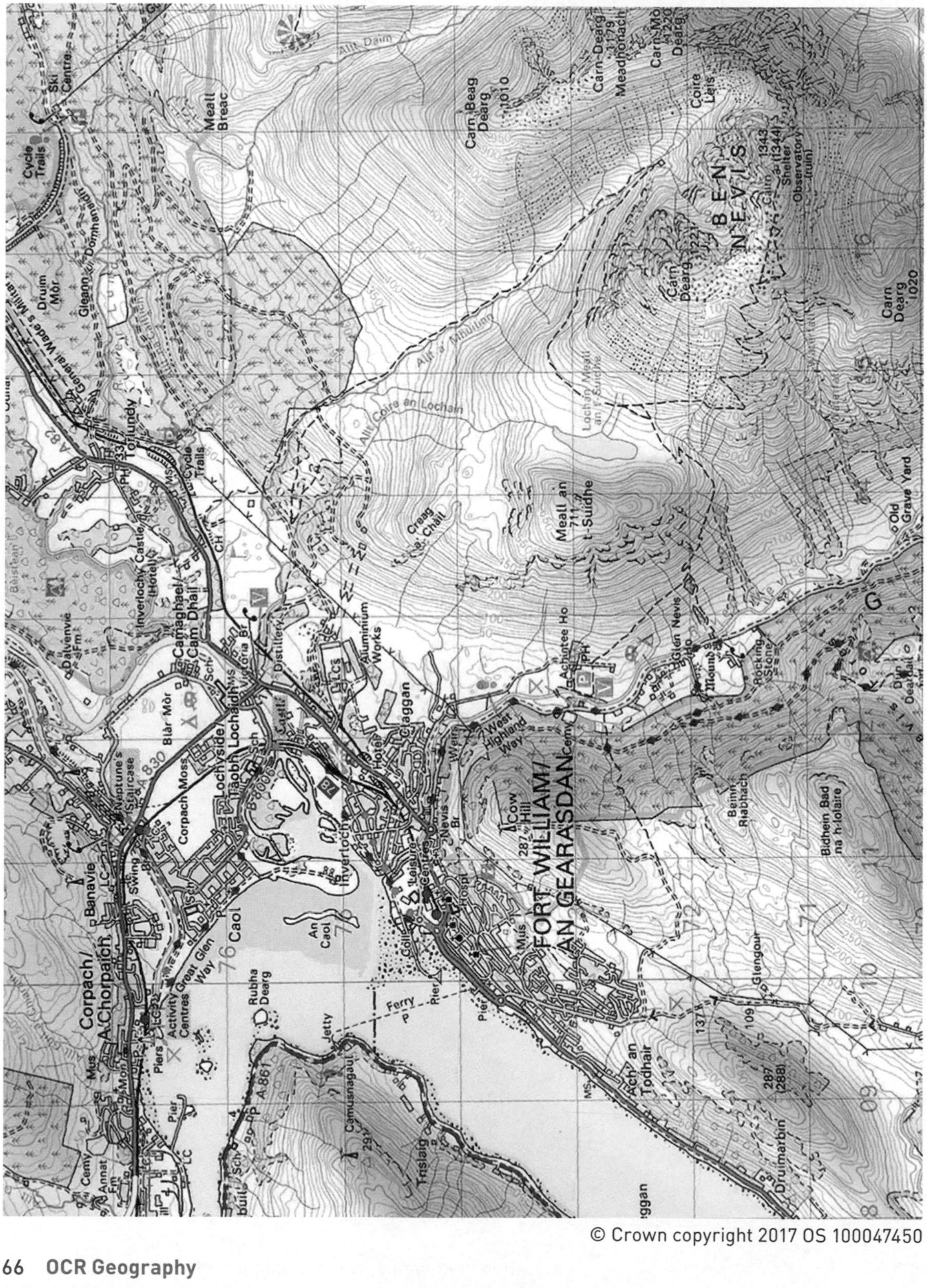

Figure 1 The Fort William area (1:50,000 OS map extract)

(a) Study Figure 1, which shows a 1:50,000 OS map extract of the Fort William area in which an AS geographical investigation is to be undertaken.

(i) State a geographical question or issue that could be investigated within the area shown in Figure 1. Justify your choice using map evidence. (4 marks)

e This question requires a clear statement of a valid geographical question or issue which could be investigated in the area of the map extract. The justification for choice of question or issue should be linked to specific evidence/features identifiable on the OS map.

Student answer

(a) (i) A geographical question for investigation in the area of the map extract could be 'How and why do people perceive the Fort William area in different ways?' a This would be an investigation of how people see, experience and understand place.

OS map evidence suggests that the investigation could be justified in this area. For example local residents are likely to identify the importance of Fort William as a service centre. b Perception of these services may vary with age. The map shows it is the largest settlement in the area and that it has a hospital, schools and churches. c The working population of Fort William and surrounding settlements would see the services and the works as places of regular employment. b An example is the distillery at 126757. c Another perception may be that of visiting tourists who see the area for recreation. b The map shows accommodation (hotel and camping and caravanning sites), communication links and other attractions such as Ben Nevis, coastal features and ski tows. c

e **4/4 marks awarded** This is a thorough answer which includes a clear statement of a valid geographical question. a The investigation is justified in three different ways b using reference to specific map evidence. c

The OS map has been chosen to elicit a wide range of possible geographical questions or issues which could be investigated. The student response above is based on a perception study, but other human or physical topics or physical/human issues could be justified in the area of this map extract. For example, these could include investigation of glaciated landforms, processes or the effects of human activity on change within glaciated landscape systems. Other possible topics could include investigation of coastal landforms, processes or the impact of human activity on this coastal landscape system. Also investigation of place identity, place representations and the impact of place-making processes are further possibilities.

Regardless of which topic is chosen, it is important to state a clear and valid question or issue that is geographical in content and that is researchable in the area shown in the map extract. The investigation should be at a scale that is practicable given the time and resources available to you/a school group at this level.

(a) (ii) Outline one method you would use to collect primary data for the investigation in (a)(i). (2 marks)

e As it is worth only 2 marks, the outline should be relatively brief. The method or technique should be clearly identified and there should be some development to show how it would be used to collect primary data.

(a) (ii) Primary data could be collected using the questionnaire technique; questions would specifically target perception of place. a Short questionnaires could be conducted with people of differing identity at various locations within the map area such as Fort William high street and visitor centres. b

e **2/2 marks awarded** This response identifies an appropriate technique for primary data collection. a There is some development of the technique with brief reference to the target population/possible locations. b

(a) (iii) With reference to Figure 1, examine the practical problems you might encounter in collecting primary data for the investigation outlined in (a)(i). (6 marks)

e This question requires identification and explanation of practical problems that might be anticipated in the collection of primary data needed for the chosen investigation in (a)(i). Clear and explicit reference to features on the OS map are required.

(a) (iii) The practical problems in collecting primary data using the questionnaire technique on perception of place could lead to unreliable data.

Ensuring adequate sample size from different localities might be a difficulty. a For example, it is possible that permission to conduct questionnaires at specific localities such as a visitor centre might not have been obtained. b There could be reluctance of local people to become involved if, for example, a relatively small place had too many interviewers, a for example in outlying settlements such as Torcastle (1378). b Finding sufficient respondents in each of the categories such as different age groups, male/female, local residents, tourists/visitors might be a problem a in relatively small settlements in the Fort William area such as Claggan, Inverlochy or Corpach (0976). b There could be insufficient visitors at certain times of the year and few local shoppers on certain days of the week. a Design of the questionnaire/questions could present problems in obtaining appropriate data on place perception or our ability to analyse the responses meaningfully. A more effective questionnaire might depend on an earlier pilot survey. a

All these problems need consideration from the outset and could be overcome by careful planning.

e 6/6 marks awarded The student shows thorough understanding of potential problems of primary data collection for the chosen investigation, and these are specifically related to the OS map area of study. Three of the five potential problems that have been identified **a** are supported by reference to a map feature or locality. **b** There is sufficient knowledge and understanding of problems of primary data collection related to the investigation outlined in part (a)(i) for this response to reach the top of Level 3.

(b) With reference to a fieldwork investigation you have carried out, evaluate the sampling strategy which you used. (12 marks)

e This question requires you to discuss the sampling strategy used in a fieldwork investigation in which you have been involved. Briefly outline the fieldwork and the required data in order to set the context. The essential requirement is that you comment on how appropriate or suitable the sampling strategy was. Evaluate the sampling strategy in terms of the type of data to be collected, the size of the parent populations, the resources and time available. Also it would be appropriate to discuss how this sampling strategy affected the reliability of the data and the opportunities to analyse them.

(b) A fieldwork investigation I have undertaken was the identification and explanation of variation in beach profiles at Sheringham, Norfolk. **a** We worked in groups of five along transect lines. Each student had a specific task in order to achieve consistent and accurate data collection. The data required included slope angles and distances for each slope facet so that we could draw profiles of the beach relief. There were measurements of particle size and roundness, and percolation rates at each sampling point. Estimates of wave height, wave frequency and angle of approach were recorded. **b**

Planning the sampling strategy was important to give reasonable coverage of the area and achieve reliable results. Different types of data were collected along the transect lines **c** working from the low water mark perpendicular to the shoreline. The tide caused a time problem but by dividing our labour we managed to complete five transects in all. **c** We selected the location of transects based on stratified sampling from our previous observations of the coastline. **c** This was appropriate since it ensured that we investigated different types of beach **d** — for example, the wide, steep shingle beach with its flat sands exposed at low tide, west of the lifeboat station, and the more varied profiles with pronounced berms between the wooden groynes further east. These landforms might have been missed if sampling had been random or systematic. **d**

The transects were about 50 metres long when we started at low tide. Along each we selected five sampling points, at regular systematic intervals of 10 metres, for our investigations of beach material and infiltration rates. **c** This large interval may have caused us to miss a particular feature such as a small berm or type of material. **d**

For size and roundness we chose 20 pebbles for measurement at each point. These were selected at random within the area of a quadrat placed on the beach. **c** Our quadrat had a grid and used generated random numbers to give each pebble an equal chance of being selected, to make our sample as reliable as possible. **d**

We had no time to repeat the measurements, but the data we did collect enabled us to attempt statistical analysis. **d** The parent population of beach sediment is very large, so our percentage sample size was not particularly representative but it was the most we could achieve in the time available. **d**

Overall our sampling design involved only a small number of transects which could not be sufficient to show all variations along this 2 km stretch of coastline. **d** Also, sampling on just one day at one time of year was another limitation in investigating beaches, which are dynamic landforms. **d** Nevertheless, there was sufficient coverage of a range of beach types, landforms and sediment variation for it to be worthwhile, especially given the number of students, time available and equipment available to us. **d**

e **12/12 marks awarded** The nature and location of the investigation is described **a** and there is an outline of the data to be collected. **b** This provides the context for the sampling strategy. Specific sampling techniques are identified **c** and evaluated in context. **d** Evaluation of the sampling strategies for the chosen investigation is comprehensive. This response achieves the top of Level 4.

Question 2

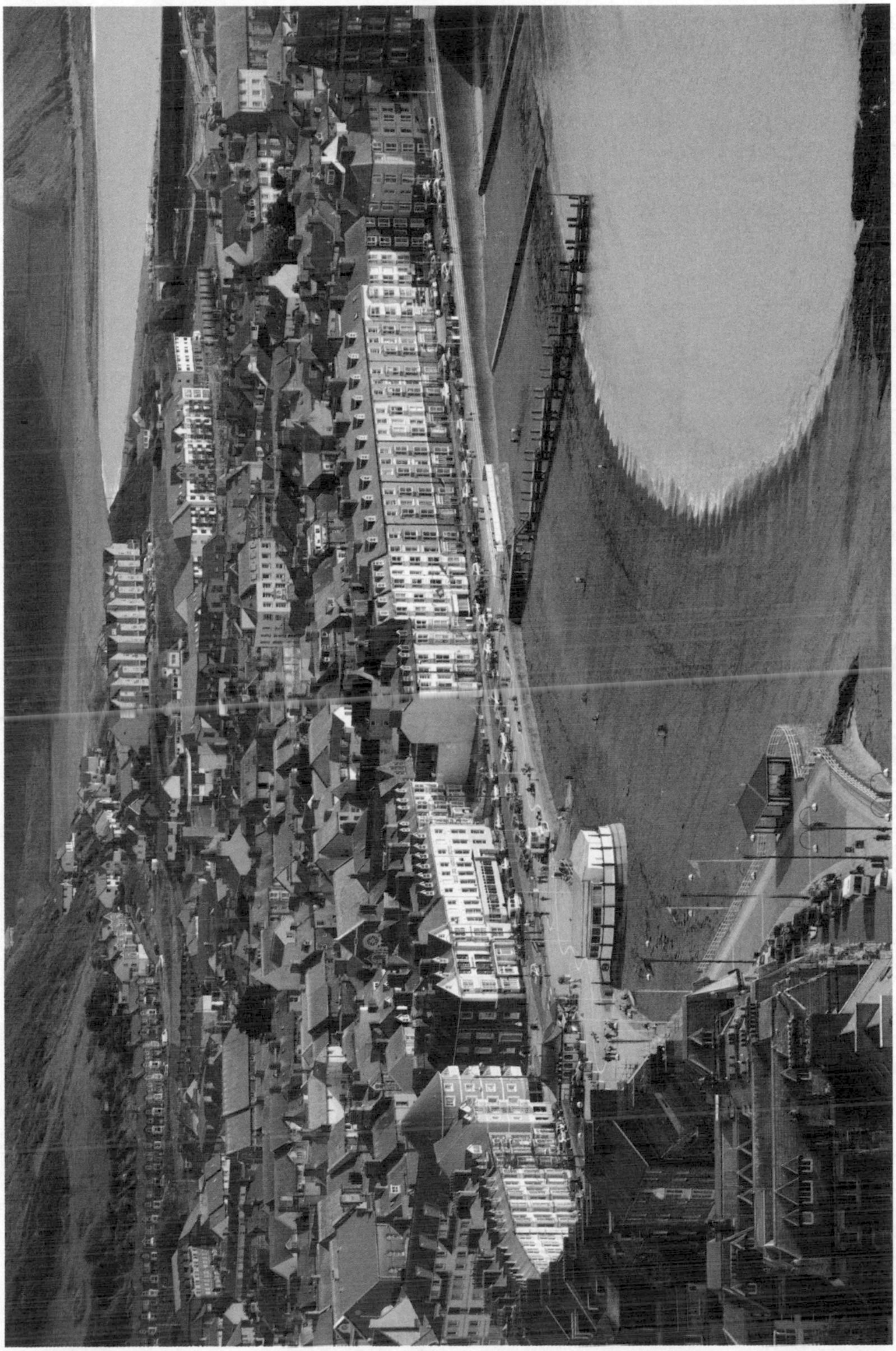

Figure 2 Aberystwyth

Study Figure 2, a photograph of Aberystwyth that shows an area in which an AS geographical investigation is to be undertaken.

(a) (i) State a geographical question or issue that could be investigated within the area shown in Figure 2. Justify your choice using evidence from the photograph. (4 marks)

e This question requires a clear and valid geographical question or issue which could be investigated through practical fieldwork in the area shown by the photograph. Choice of question must be justified with reference to specific evidence in the photograph. This could involve named features, landforms, scales and distances, for example, and/or practical considerations such as access, safety and time.

Student answer

(a) (i) It would be possible to investigate the geographical question, 'How and why does the size of beach sediment particles vary with distance from the sea wall at Aberystwyth?' **a** This is based on the premise that contrasts in energy between the stronger swash and weaker backwash of waves causes sorting of beach material. Using cars and people to give an idea of scale, the photograph shows that beach width is sufficiently large **b** for significant variation in particle size. The height of beach sediment accumulation against the sea wall towards the foreground of the photograph **b** suggests there has been transport of beach material by high energy waves and the probability of sorting by size. **c** The sea wall **b** is a fixed point from which measurements could be taken. **d**

e **4/4 marks awarded** There is a clear and correct statement of an appropriate geographical question. **a** In asking 'how' and 'why' the question requires both identification and explanation of any variation in beach material size. The question has been justified in this specific area by linking evidence in the photograph **b** to either knowledge of landform and process **c** or practical consideration. **d**

The photograph allows a range of possible options for choice of a geographical question or issue. While a physical topic on a coastal landform has been chosen in the response above it would be equally valid to cite an alternative physical investigation or a human topic or physical/human issue. For example, features of the photograph could prompt investigation of another physical topic such as the impact of coastal management on change within this system. Another option could involve how people perceive Aberystwyth in different ways. It would be possible in this town to investigate spatial patterns of social inequalities at local scale or the impact of economic change or placemaking.

Whichever topic you choose, you should ensure that the question or issue is geographical in content and that it would be possible to collect appropriate data in the area of the resource. The investigation should be at a scale which is practicable given the time and resources available to you/a school group at this level.

(a) (ii) Identify two sets of primary data required for the investigation in (a)(i). (2 marks)

ⓔ You should identify two different types of primary data specifically required for the investigation stated in (a)(i).

> **(a) (ii)** Two different types of primary data which could be collected to investigate the question in (a)(i) are measurements of the distance of each sampling point from the sea wall a and the length or a-axis of a number of beach sediment particles at each sampling point. a

ⓔ **2/2 marks awarded** Two appropriate types of primary data a are clearly stated which are of direct relevance to answering the question identified in (a)(i). Examples of other data which might help in the explanation could include measurement of infiltration rates on the beach and estimates of wave energy and wave characteristics.

(a) (iii) With reference to Figure 2, explain how you would ensure that the data collected for the investigation in (a)(i) was reliable and accurate. (6 marks)

ⓔ An essential part of planning an investigation is to ensure that the data collected is as reliable and accurate as possible. The focus of your answer should be on practical ways you would reduce the risk of unreliable or inaccurate data. There should be explicit reference and links to relevant features shown in the photograph.

> **(a) (iii)** The risk of collecting data which are unreliable would be minimised by use of an appropriate sampling strategy. A straight line transect a would be possible across the beach from sea wall to water level. b The position of this line is likely to show variation in sediment size which has been sorted by wave energy up and down the beach. c
>
> A number of sampling points placed systematically at frequent intervals a along the transect line would also ensure more reliable data. This should be possible since the beach appears to be wide at low tide. b The greater the number of sampling points the more reliable the analysis if using a scatter graph to represent possible correlation. c
>
> Measurement of the a-axis of a large number of beach sediment particles selected at each sampling point a would also ensure the data was more reliable.
>
> The risk of inaccuracy could be minimised by use of accurate instruments such as callipers a, a second person checking the reading a and by careful recording of the data in the field a on a pre-prepared spread sheet. Also a useful fixed point from which to take distance measurements down the beach a is provided by the sea wall. b
>
> Consulting the times of high and low tide a could provide useful information on time available for collection of the beach data.

e 6/6 marks awarded This response includes a number of valid suggestions for reducing the risk of collecting data which are not reliable or accurate. [a] Where possible these techniques have been linked to photographic evidence. [b] Explanation of the influence of the fieldwork techniques on reliability and accuracy is offered for the first two points. [c] The latter part of the response includes valid techniques but they are undeveloped in the context of the question either without reference to the photograph or lacking explanation. The response reaches the top of Level 3.

(b) With reference to a fieldwork investigation you have carried out, assess the relative value of the primary and secondary data collected. (12 marks)

e This question requires discussion of the relative value of the primary and secondary data collected in a fieldwork investigation in which you have been involved. The contributions and usefulness of the primary and secondary data collected should be examined in the context of a specific investigation.

(b) My fieldwork investigation in human geography was on the topic of social inequality in London. We selected two areas, part of Shadwell in Tower Hamlets, an inner urban area, and part of Hampstead Garden Suburb, an outer urban area in Barnet. Our areas for investigation were defined by the OS maps published for Lower Layer Super Output Areas in the last Census. The aim of the fieldwork was to identify and explain contrasts in the socio-economic characteristics of the two areas. [a]

It was decided firstly to assess the social inequalities by fieldwork observations. Primary data were collected through housing surveys. These included the type and size of house, number of garages if any and the size of the plot. [b] We recorded these indices using self-generated data recording sheets. Also the characteristics and quality of the surrounding environment were observed [b] and recorded. Using a combination of fieldwork and OS maps we also measured access to public transport, including number of bus stops, train/tube/DLR stations in each area and location of services such as GPs, schools and retail. [b]

These primary data were useful in giving an overall contrast in social inequality but there were other aspects which we could not observe or which were too personal to consider using a questionnaire. Some of the data depended on our own judgements and may not have been accurate, such as the size of a particular dwelling. [d] The data we could observe only gave part of the socio-economic characteristics of each area. For example there were limitations in our attempts to estimate an idea of wealth. We could not observe with any reliability other population characteristics such as age or ethnicity. Field observations gave no idea of qualifications, employment, health and benefits. [d]

Use of secondary data therefore became very important in adding to our investigation. The information for these areas was available in the online national statistics for the 2011 Census. The statistics we used were various socio-economic indices for Lower Layer Super Output Areas 025C, Tower Hamlets, and 033B, Barnet. [c]

> These secondary data were very useful since we selected socio-economic indices for our two areas which we could not otherwise observe by fieldwork. They included population density, persons per room, country of birth, car ownership, housing tenure, 16–74 year olds with no qualifications and those with limiting long-term illness. **d** This added to our fieldwork observations. It enabled comparison between the two areas and demonstrated clear inequalities. However, the data were six years out of date and did not indicate the changes that are likely to have occurred since the last Census. **d**
>
> There were advantages and limitations in using both the primary and the secondary data. They both contributed to the identification and understanding of urban social inequalities in different ways.

e 12/12 marks awarded This response offers a brief but useful outline of the fieldwork investigation. This is important in order to place the evaluation of primary and secondary data in context of the investigation. **a** The primary data are identified **b** although this section on the nature of the fieldwork and the types of data collected lacks detail and is not convincing. The source of secondary data is also stated **c** and the contribution to the investigation is discussed, although reference to secondary data is confined to census data alone. Both primary and secondary data are evaluated in the context of the investigation with reference to some advantages and some limitations of each. **d** Overall this response achieves a mark at the top of Level 4 since the value of the two types of data is discussed with well-developed arguments.

Knowledge check answers

1 The data you collect yourself in the field are primary data, as are some published data. The important aspect is that they are unprocessed, raw data. Once data have been processed in some way (analysed/interpreted) they become secondary data.

2 Quantitative approaches are based on numerical data and nearly always involve statistical testing. Qualitative approaches focus on how the world is viewed, experienced and made by people. Often, interviews or direct observation of people are used. In addition, patterns and processes are investigated by analysing text, images or music.

3 Direct observation of people, interviews with an individual or group and textual analysis, which includes sampling text, visual or musical material.

4 Coding material helps you sift through what can be a mass of apparently unconnected material, organising the data so that you can make a rigorous and meaningful analysis. Coding helps you be systematic as you go through material such as interviews.

5 Non-spatial sampling helps you collect data from a population so that it is representative of the population as a whole. Spatial sampling aims to achieve the same but, when location is important, it ensures that you survey a representative sample of locations.

6 Crowd-sourced data rely on people volunteering to submit information online. The people and their data may not be a representative sample.

7 Scale, key, title and north point are the four elements that all your maps must include.

8 A label is a concise and precise identification of a feature. An annotation gives more detail and includes comments.

9 Using percentages are useful as you are then comparing equivalent data as the influence of the total size of the data set (absolute numbers) is minimised.

10 The dependent variable is affected by a change in the independent variable. The independent variable is expected to cause change in the dependent variable.

11 Testing the significance of a statistical result allows you to assess how likely it is that the result might have occurred purely 'by chance'. If chance can be eliminated, then you can proceed with your analysis and conclusions confident that relationships exist in your data.

Note: **Bold** page numbers indicate key terms.